PORPHYRY

To Gaurus On How Embryos are Ensouled

and

On What is in Our Power

PORPHYRY

To Gaurus On How Embryos are Ensouled
and
On What is in Our Power

Translated by
James Wilberding

B L O O M S B U R Y
LONDON • NEW DELHI • NEW YORK • SYDNEY

Bloomsbury Academic
An imprint of Bloomsbury Publishing Plc

50 Bedford Square
London
WC1B 3DP
UK

1385 Broadway
New York
NY 10018
USA

www.bloomsbury.com

Bloomsbury is a registered trade mark of Bloomsbury Publishing Plc

First published in 2011
Paperback edition first published 2014

British Library Cataloguing-in-Publication Data
A catalogue record for this book is available from the British Library.

ISBN HB: 978-0-7156-3858-3
PB: 978-1-4725-5787-2
ePDF: 978-1-4725-0167-7

Library of Congress Cataloging-in-Publication Data
A catalog record for this book is available from the Library of Congress.

Acknowledgements

The present translations have been made possible by generous and imaginative funding from the following sources: the National Endowment for the Humanities, Division of Research Programs, an independent federal agency of the USA; the Leverhulme Trust; the British Academy; the Jowett Copyright Trustees; the Royal Society (UK); Centro Internazionale A. Beltrame di Storia dello Spazio e del Tempo (Padua); Mario Mignucci; Liverpool University; the Leventis Foundation; the Arts and Humanities Research Council; Gresham College; the Esmée Fairbairn Charitable Trust; the Henry Brown Trust; Mr and Mrs N. Egon; the Netherlands Organisation for Scientific Research (NWO/GW); the Ashdown Trust; Dr Victoria Solomonides, the Cultural Attaché of the Greek Embassy in London.
The editor wishes to thank John Dillon, Vivian Nutton, and Andrew Smith for their comments, Sarah Francis for preparing the volume for press, and Deborah Blake at Duckworth, who has been the publisher responsible for every volume since the first.

Typeset by Ray Davies.

Contents

Preface

My interest in the *Ad Gaurum* was kindled years ago as I was working on the first draft of the commentary on Plotinus' *Ennead* 2.1 that was ultimately published as *Plotinus' Cosmology*. In 2.1.5 Plotinus appeals to the different souls responsible for the formation of human bodies and celestial bodies in order to explain why the latter but not the former are everlasting. As is often the case in Plotinus, the passage seemed at least *prima facie* ambiguous regarding the identity of the soul responsible for human formation: Was it the individual lower soul of the offspring, that of the parent, or the World-Soul? All three seemed to be reasonable candidates, and indeed I soon discovered that this is a question that was much discussed by Platonists (cf. Galen *On the Formation of Fetuses* 699,3ff.), which should come as no surprise given the complexity and scope of soul in the Neoplatonic universe. And so I was very grateful when I discovered that the *Ad Gaurum*, a treatise most probably to be attributed to Plotinus' student Porphyry, supplied a considered and carefully articulated response to this question along with much else of value to those interested in Neoplatonic psychology. The translation of *On What is in Our Power* has been added to complete the volume, since both approach the issue of how and why a particular soul enters into a particular body, albeit from very different starting points. These are two important treatises by Porphyry that are appearing in English now for the first time.

I would like to thank Williams College and the Alexander von Humboldt Foundation for funding research leaves during which most of the work on this translation was completed, and specifically Christoph Horn and Andreas Speer for generously welcoming me into the University of Bonn and the Thomas Institute in Cologne respectively. I also owe a number of colleagues and friends my warmest thanks. Ian Mueller, Andrew Smith, Philip van der Eijk and an anonymous vetter all took time to read and comment on the translation and introduction of the *Ad Gaurum*, and F. Gavin Hardy, Dominic O'Meara and Jan Opsomer kindly discussed individual passages or problems of the *Ad Gaurum* with me. Wolfgang Hübner and Dorian Gieseler Greenbaum graciously agreed to check over the astrological details in my translation of *On What is in Our Power*, and I received additional help from Christian Helmig and the anonymous vetter for the series. Two of my students, Elaine Watts and Vanessa Hollaway, deserve thanks for their efforts and patience with this rather difficult text in a Greek seminar. I am grateful also to Richard Sorabji for making this work possible as well as providing valuable input. I owe thanks and apologies to Fiona Leigh for her patience and effort on a project that ultimately did not reach its completion during her tenure, and finally to Sarah Francis for all the subsequent editing. The final product has surely profited greatly from all of these contributions, but for the errors that doubtless remain I am alone responsible.

Just prior to receiving the proofs for this volume I learned of the untimely death of my PhD supervisor and mentor Ian Mueller, and I would like to dedicate this volume to his memory in gratitude for all he has done for me.

Conventions

[ ] Square brackets indicate content that is only implicitly contained in the text but that is desirable to make more explicit in the translation.

< > Angle brackets indicate additions to the Greek text, only where these are not already included in the body of Kalbfleisch's printed text.

{ } Curly brackets indicate restorations in the text, only where Kalbfleisch's suggested restoration is not followed.

… A series of dots indicates text that Kalbfleisch has been unable to restore.

Abbreviations

While an attempt has been made to use only English titles in the Introduction and Translation, the following standard abbreviations have been used in the notes.

Aëtius
Plac. = *On the Opinions of the Philosophers*
Alcinous
Didask. = *The Handbook of Platonism*
Alexander of Aphrodisias
de An. = *On the Soul*
Ethic. Prob. = *Ethical Problems*
in Metaph. = *On Aristotle's Metaphysics*
in Meteor. = *On Aristotle's Meteorology*
in Top. = *On Aristotle's Topics*
Aristotle
DA = *On the Soul*
EN = *Nicomachean Ethics*
GA = *On the Generation of Animals*
HA = *History of Animals*
Metaph. = *Metaphysics*
Phys. = *Physics*
PN = *Parva Naturalia*
Rhet. = *Rhetoric*
Asclepius
in Metaph. = *On Aristotle's Metaphysics*
Augustine
de Trin. = *On the Trinity*
Damascius
de Princ. = *On the First Principles*
Galen
de Demonst. = *On Demonstration*
de Foet. Form. = *On the Formation of Fetuses*
de Hist. Phil. = *On the History of Philosophy* (spurious)
de Mar. = *On Marasmus*
de Mus. Dissect. = *On the Anatomy of Muscles*
de Plac. Hipp. et Plat. = *On the Doctrines of Hippocrates and Plato*
de Prop. Plac. = *On My Own Opinions*

de Sem. = *On the Seed*
de Ther. ad Pis. = *To Piso on Theriac*
de Usu Part. = *On the Usefulness of the Parts of the Body*
in Hipp. Epid. 2 comment. = *Commentary on Hippocrates' Epidemics II*
in Hipp. Epid. 6 comment. = *Commentary on Hippocrates' Epidemics VI*
Inst. Log. = *Introduction to Logic*

Hermeias
in Phaedr. = *Scholia on Plato's Phaedrus*

Hippocratic Corpus
Alim. = *On Nutriment*
Mul. = *On the Diseases of Women*
Nat. Puer. = *On the Nature of the Child*
Superf. = *On Superfetations*
Vict. = *On Regimen*

Iamblichus
De an. = *On the Soul*
De comm. math. = *On the General Mathematical Science*
De myst. = *On the Mysteries*
Protr. = *Protrepticus*
Vita Pyth. = *Life of Pythagoras*

Marcus Aurelius
Med. = *Meditations*

Michael of Ephesus
in GA = *On Aristotle's Generation of Animals*
in SE = *On Aristotle's Sophistical Refutations*

Nemesius
de Nat. Hom. = *On the Nature of Man*

Olympiodorus
in Meteor.= *On Aristotle's Meteorology*

Oribasius
Coll. Med. = *Medical Collections*

Philo
de Somn. = *On Dreams*

Philoponus
Contra Proclum = *Against Proclus on the Eternity of the World*
de Opif. Mundi = *On the Creation of the World*
in Cat. = *On Aristotle's Categories*
in DA = *On Aristotle's De Anima*
in Phys. = *On Aristotle's Physics*

Plato
Alc. 1 = *Alcibiades I*
Ep. 7 = *Seventh Letter*
Epin. = *Epinomis*
Lach. = *Laches*

Leg. = *Laws*
Phdr. = *Phaedrus*
Phil. = *Philebus*
Rep. = *Republic*
Symp. = *Symposium*
Theaet.= *Theaetetus*
Tim. = *Timaeus*

Pliny
Nat. Hist. = *Natural History*

Plotinus
Enn. = *Enneads*

Plutarch
de Plac. Phil. = *On the Opinions of the Philosophers* (spurious)

Porphyry
AG = *Ad Gaurum*
de Abst. = *On Abstinence from Killing Animals*
de Antro Nymph. = *On the Cave of the Nymphs*
de Phil. ex Orac. = *On the Philosophy from Oracles*
de Regr. An. = *On the Return of the Soul*
de Styge = *On Styx*
in Ptol. Harm. = *On Ptolemy's Harmonics*
Marc. = *Letter to Marcella*
Quest. Homer. = *Homeric Questions*
Sent. = *Sentences*
Summ. Zêtêm. = *Miscellaneous Questions*
VP = *Life of Plotinus*

Proclus
de Sacr. = *On Sacrifice*
Elem. Theol. = *Elements of Theology*
in Parm. = *On Plato's Parmenides*
in Remp. = *On Plato's Republic*
in Tim. = *On Plato's Timaeus*
Theol. Plat. = *Platonic Theology*

Sextus Empiricus
AM = *Against the Mathematicians*

Simplicius
in Phys. = *On Aristotle's Physics*

Soranus
Gyn. = *Gynecology*

Stoics
LS = Long & Sedley (see bibliography)
SVF = von Arnim, *Stoicorum Veterum Fragmenta*

Syrianus
in Metaph. = *On Aristotle's Metaphysics*

Tertullian
de An. = *On the Soul*

Themistius
in DA = *Paraphrase of Aristotle's On the Soul*
in Phys. = *Paraphrase of Aristotle's Physics*
Theophrastus
de Caus. Plant. = *On the Causes of Plants*
Hist. Plant. = *Research on Plants*

PORPHYRY

To Gaurus On How Embryos are Ensouled

Textual Emendations and Conjectures

1.3.5-6 [34,1-2 K] Reading *kath' ho proelthontôn enesti* with Festugière for *kai † proselthontôn enesti*

2.4.4 [35,25 K] Reading *hama* with Kroll for *ara*

3.4.13 [37,14 K] Reading *têi* for *tais* as Kalbfleisch suggests in the critical apparatus

4.7.4 [39,22 K] Reading *ouk esti* with Kroll for *ouketi*

4.8.9-10 [40,7-8 K] Reading *hôs heterôs legomenês kai plêgais kai pathesi hupokeimenês psukhês* for *hôs heterôs kai plêgais kai pathesi legomenês psukhês* as Kalbfleisch suggests in the critical apparatus.

4.10.1-3 [40,20-2 K] Reading *aphantastous <ou> tais kriseis tautês tês doxastikês psukhês sunarmosas tês gnôristikês* for *aphantastous † tas pros tês kat' autês tês doxastikês psukhês sunarmosas tou gnôrismatos*

5.3.5 [41,17 K] Reading *sunendidomenês* for *sunekdidomenês* as Kalbfleisch suggests in the critical apparatus

6.2.3-4 [42,19-20 K] Reading *ousias apo tôn gegennêkotôn* with Mynas for *ousias † axia tôn gegennêkotôn*

6.2.9 [42,25 K] Reading *amoiros* with MS and Deuse for Kalbfleisch's *emmoiros*

6.2.10 [42,26 K] Reading *legetai* with Festugière (and as in text) for Kalbfleisch's *agetai* (in the *Nachträge*)

8.3.4 [44,25 K] Reading *para* with Plato *Tim.* 91C3 for *peri*

9.1.5 [45,9 K] Reading *phutikê <hê> dioikêsis* as Kalbfleisch suggests in the critical apparatus

9.2.6 [45,15 K] Reading *ekthrepsôntai* for *ekthrepsôsin*

10.1.7 [46,18 K] Reading <*tês*> before *en* as Kalbfleisch suggests in the critical apparatus

10.3.4 [46,27 K] Reading *apotelousês* for *apotelousin* as Kalbfleisch suggests in the *Nachträge*

10.5.6 [47,21 K] Reading *ta <mê> krathenta* with Deuse for *ta krathenta*

10.6.3 [47,28 K] Reading *mê<keti êi> kata tên* for *mê † kat' autên tên* as Kalbfleisch suggests in the critical apparatus

11.2.10 [48,26 K] Reading *ophtheis athroon* for *haphtheis haphthenti*

11.2.13 [49,1 K] Inserting *ho êlektros* before *ta karphê*

11.3.17 [49,20 K] Reading *akhronos* with Festugière for Kalbfleisch's *akhr.i*

12.2.1-2 [50,11-12 K] Reading *ouk hôs atopon apodexamenos* for *ouk eis atopon anadeixamenos*

12.3.8-9 [50,22-3 K] Reading *kai ou deos mê tis memphêtai to tôn paidiôn aphaireisthai* for *kai † oude osmê tis memph* * * *polôn aphaireisthai*, following suggestions by Kalbfleisch and Kroll.

13.1.6 [52,8 K] Reading *kata to eidos <kai> êremon monon* for *kata to eidos êremon men ên*

14.2.2 [54,1 K] Reading *ouden propherousi biastikon* for *ouden diapherousi biastikôs* as Kalbfleisch suggests in the critical apparatus

14.3.1 [54,4 K] Reading *alla* for *ara* as Kalbfleisch suggests in the critical apparatus

14.4.8 [54,22 K] Reading *oudamou tôn gnôristikôn* for *oudamou tôn gig...rist...*

16.2.2 [56,6 K] Reading *eugonia* for *eutonia* as Kalbfleisch suggests in the critical apparatus

16.2.2 [56,6 K] Reading *tês mellousês* with Diels for *têi mellousêi*

16.2.3 [56,7 K] Reading *ekeino* for † *kenoi* as Kalbfleisch suggests in the critical apparatus

16.2.5 [56,9 K] Reading *hôs eugonia phuseôs kai aretê psukhês* for *hôs eugonia phuseôs aretê psukhês † kai aretê phusis* as Kalbfleisch suggests in the critical apparatus

16.2.10 [56,14 K] Reading *hês idion hê* for *hês † idrê* as Kalbfleisch suggests in the critical apparatus

16.4.3 [56,23 K] Reading *hotan apo tês mêtros mêketi, mêde to em{bruon} êi kata tên* with Festugière for *hotan † apo tês mêtros mêketi to en hôi kata tên*

16.5.4 [57,2 K] Reading *hêtis an <êi> k{ata kairo}n psukhê* with Festugière for *hêtis an k......n psukhê*

16.7.4-5 [58,3-4 K] Reading *dêloi d' horôn ho a{narmostos} hôs {heterophthalmos ho}ti thaterôi parên hê harmonia* for *dêloi d' horôn ho a......... hôs .... ho}ti thaterôi parên hê harmonia*

16.9.1 [58,11 K] Reading *ekhôristhê* with Diels for *ekhôrê(sen)*

17.2.8-9 [58,30-59,1 K] Reading *prôta de ta dunamena e{pikourein t}ous hu{per} t{ou Pla}tônos {makhom}enous t{ithemen}* for *prôta de ta dunamena e{lenkhein t}ous hu... t{ou Pla}tônos .....enous t{ithemen}*

17.5.1-3 [59,19-21 K] Reading *{hôsper oun atop}os ho tês thruallidos tên ek puros {exapsin} exôthen phaskôn gegonenai, {hôsper hê dia tês tôn li}thôn s{un}krouseôs exôthen ên, houtôs kai ho...* for *{hôsper oun ouk atop}os ho tês thruallidos tên ek puros {exapsin} exôthen* *phaskôn gegonenai, {epeidê ek tês tôn li}thôn s{un}krouseôs exôthen ên, houtôs kai ho...*

17.5.7-8 [59,25-6 K] Reading *epi skô{lêkôn kai eulôn kai muiôn kai ôiôn hupê}nemiôn* for *epi skô{lêkôn kai eulôn kai ....nemôn.* The final *ôiôn hupê}nemôn* is suggested by Kalbfleisch in the critical apparatus.

17.6.13-7.1 [60,11-12K] Reading *tôn {karpôn monon, alla kai tôn zôiôn, hênôntai en sper}ma{si} kai husteron diakrinontai* with Festugière for *tôn ....{kal}am{ês} kai husteron diakrinontai*

17.7.8-9 [60,19-20 K] Reading *ana{plattesthai kata diaphorous kairous e}isionta* with Festugière for *ana{plattesthai ....{e}isionta*

17.7.13 [60,24 K] Reading *sunarmozein* for *sunomartein*

Introduction

1. The authorship of *Ad Gaurum*

The *Ad Gaurum* was discovered in the mid-nineteenth century as part of the remains of a thirteenth-century manuscript. As Karl Kalbfleisch, who in 1895 published the edition of the *Ad Gaurum* text on which this translation is based, explains in his introduction, the first page of these remains contained bits of the conclusion to Galen's *On Marasmus,*[1] followed by two texts, Galen's *Introduction to Logic*[2] and the *Ad Gaurum*.

In the manuscript the *Ad Gaurum* is itself also attributed to Galen, the full title being *Galenou pros Gauron peri tou pôs empsukhoutai ta embrua*, but as Kalbfleisch has convincingly argued, this attribution cannot be right. Just how it came to be attributed to Galen is unclear. Kalbfleisch suggests that at some point the author's name vanished from the treatise, and since the *AG* was certainly read in medical circles, this authorial gap was filled with the name of the physician *par excellence*, namely Galen.[3] Be that as it may, he supplies convincing arguments designed to show that the doctrine advanced in the *Ad Gaurum* differs considerably from the doctrine to be found in Galen's writings on embryology, namely *On the Formation of Fetuses,*[4] *On the Seed,*[5] and the Books 14 and 15 of *On the Usefulness of the Parts of the Body*.[6] In these treatises we find Galen presenting views that are in direct opposition to some of the main conclusions of *Ad Gaurum*:

1. The fetus is said to live in the manner of an animal as soon as its heart starts beating.[7] Likewise, in *On the Usefulness of the Parts of the Body*[8] Galen says that in *On Demonstration* and *On the Doctrines of Hippocrates and Plato* he has shown that the fetus acquires animal status as soon as its parts are formed.[9] This contradicts the main thesis of *Ad Gaurum*, that the offspring becomes an animal only at birth. To elaborate on Kalbfleisch's point: Galen and Porphyry agree that the beginning of animal life comes with the completion of the organs,[10] since otherwise nature would have produced something in vain – organs without the principle to operate them. Yet Porphyry insists that the organs are not *fully* completed until birth, and so he can say that the fetus does not have a head or heart or even a liver yet.[11] Galen, by contrast, takes a more gradualist

approach. Although Galen seems to have developed different views on the order of development of the embryo,[12] his considered position was that the liver is fully developed first,[13] followed by the heart and then the brain. This position was largely motivated by the Platonic view that the three parts of soul reside in these three parts. Thus, the early development of the liver corresponds to the vegetative or appetitive soul's foundational role,[14] and in this stage it is governed like a plant.[15] Once the heart is formed and starts to beat, it begins to live in the manner of an animal – though this is qualified by specifying the kind of animals it resembles, namely mussels, trumpet-shells, pinnae, oysters and shell fish.[16] In short, though it no longer lives like a plant, it is not yet a full-fledged animal, but rather resembles what have been called 'fuzzy natures' that occupy the logical space between plant and animal.[17] The brain comes very late, when facial features are formed, and the powers connected to it (sight, hearing, taste, smell, touch, imagination, ratiocination and memory) seem to emerge and become more potent gradually (with this process continuing even after birth).[18] This gradualist approach seems closer to Iamblichus' view than to Porphyry's.[19]

2. In *On the Usefulness of the Parts of the Body*[20] Galen criticizes physicians who want to deny that fetuses move by means of impulse; attributing them to the womb is simply wrong.[21] But in *Ad Gaurum* 7.1-8.4 it is argued that no motion of the fetus is due to impulse but rather to the womb.

3. Galen takes an agnostic position with respect to the question of which soul is responsible for forming the embryo, whether it is a parental soul, or nature, or the World-Soul.[22] This is an illustration of Galen's general tendency to avoid taking sides on transcendental issues that lack a solid empirical foundation. But in the *Ad Gaurum* we encounter very specific views on this issue. It is the offspring's nature that is responsible for constructing the body with the help of its parents.[23]

To these we might add:

4. Galen is a vocal proponent of the two-seed theory – both father and mother supply seed,[24] but there is no mention of a female seed in the *Ad Gaurum*.[25] This leads to further disagreements, e.g. Galen's explanation of maternal resemblance is based on his doctrine of the female seed, whereas the *Ad Gaurum* appears to explain the mother's contribution to the child's appearance through her role as psychological captain during conception and gestation.[26]

While it is possible that Galen had changed his mind on these issues, this is unlikely, as the author of the *Ad Gaurum* does not signal that he previously held a different position nor is there any indication that the author has written other embryological treatises. However it came to be attributed to Galen, it would seem that Galen cannot be the author.[27]

Kalbfleisch also argues very persuasively that in fact Porphyry is the author of our text. A number of philosophical details indicate that the author must be a Neoplatonist of some stripe. This is particularly clear in the discussion of how the mother's power of representation can affect the offspring's nature (*Ad Gaurum* 6.2-3) where we find the ontological chain of intellect, discursive reason, the irrational soul, and nature described in the customary terms of generation and reversion.[28] He notes that the style of the *Ad Gaurum* is certainly unlike Plotinus' characteristic breviloquence; nor, he suggests, does it display the 'boring breadth' of Iamblichus' prose.[29] But a number of parallels between the *Ad Gaurum* and the documented contents of Porphyry's work provide some solid reasons for taking him to be the author:

1. The *Ad Gaurum*'s textbook claim that every generator generates something worse than itself (*AG* 14.3.8-9 [54,11-12 K]) can be found again in Porphyry *Sent.* 13.

2. The author of *Ad Gaurum* says that *daimons* are unique in their ability to change their own physical appearances via their representations (*AG* 6.1). This can also be found in Porphyry, e.g. *On the Cave of the Nymphs* 11.13-17 and *Sentences* 29.[30]

3. Porphyry's solution to the Body-Soul problem – a problem in which he was particularly interested (*Life of Plotinus* 13) – emphasizes that the soul's presence cannot be bodily (or spatial), nor can it be a kind of mixture (*Sentences* 33.49-53), being rather based on likeness and suitability (*Sentences* 29.21-2; 37.41-4; 38.11) such that the body, as it were, draws the soul to it (*On Abstinence from Killing Animals* 2.48.1-2) in such a way that no obstacle can keep it from being there (*On Abstinence from Killing Animals* 2.47.10-11; *Sentences* 28). All of this is found again in the *Ad Gaurum* with striking linguistic echoes (11.2.12-14 [= 48,28-49,2 K]; 13.7.1-5 [= 53,17-21 K]; 14.4.8-11 [= 54,22-5 K]).

4. The *Ad Gaurum* similarly explains the relationship between the individual soul's presence to Intellect in terms of suitability and likeness (*Ad Gaurum* 12.3), and this too can be found in Porphyry 'in very similar expressions' (*Sentences* 32.51-5; 35.5-7; 40.29-31; *On Abstinence from Killing Animals* 1.29.10-15).

5. The *Ad Gaurum*'s account of sense perception is also based on likeness and suitability and rejects more physical or material explanations (*AG* 11.2). This fits exactly with Nemesius' brief report (*On the Nature of Man* 59,13ff. Morani) of Porphyry's now lost *On Perception*.

6. At one point (10.5.5-6.3 [= 47,20-8 K]) the *Ad Gaurum* appropriates the Stoic doctrine of complete blending in order to explain how two psychic entities can form a unity without losing their identity. Here we are referred to 'other sacred books' where the author goes into greater detail on complete blending. Kalbfleisch sees this as a

reference to the *Miscellaneous Questions* (Nemesius *On the Nature of Man* 43.1-11 Morani) where Porphyry does indeed discuss blending and unity.[31]

7. Iamblichus (*On the Soul* §31 Finamore-Dillon) reports that Porphyry held precisely the view argued here, namely that the soul enters at birth.

8. Kalbfleisch's 'keystone' argument is that Michael Psellus reports that Porphyry, Hippocrates and Galen each wrote a treatise on whether the embryo was an animal (Porphyry Fr. 267 [Smith]) in which it is argued that the embryo is not an animal and that it is moved by nature rather than by a soul.[32]

9. Finally, some incidental evidence is provided by the anonymous Christian author of *Hermippus* (*On Astrology*) who launches an attack against what is obviously the *Ad Gaurum*. This is perhaps to be expected if Porphyry was indeed the author.

The content of some of these parallels is certainly common to any number of Neoplatonists. (1), for example, is just as easily found in Plotinus,[33] Iamblichus[34] and Proclus.[35] Likewise explaining the presence of the intelligible to the sensible in terms of suitability is common to Neoplatonists from Plotinus onwards,[36] as is the likeness- and sympathy-based account of perception.[37] Nevertheless, the linguistic similarities between the Porphyry and the *Ad Gaurum* passages are striking. Although Plotinus's views on sensation, for example, are similar to those found in the *Ad Gaurum*, he never discusses a 'cone' (*kônos*) extending from the eye as the *Ad Gaurum* does, but it is precisely this term that reappears in Nemesius' summary of Porphyry's *On Perception*. Parallels of this sort, together with points (6)-(8), are the reasons behind the virtually universal consensus that Porphyry is indeed the author of the *Ad Gaurum*. Nor should the fact that the *Ad Gaurum* nowhere refers to Plotinus by name count against Porphyry's authorship. Although it might be reasonable to expect Porphyry to refer to his teacher when writing on philosophical matters, Porphyry for whatever reason does not always do so.[38]

2. Which soul is responsible for creating the embryo?

The question of agency in the generation of the embryo was much discussed in antiquity,[39] but the psychological complexity of the Neoplatonic cosmos adds a further level of difficulty to this question. Either or both of the parents could be responsible, and since each of the parents has both a higher (descended) and lower soul, it is reasonable to ask of each of these four souls what role if any it is playing in reproduction. Moreover, one might think that the offspring's own individual soul is itself responsible for the production of its own body by means of its lower soul. Or else one might think that it is the World-Soul that is primarily responsible for the generation of the bodies of all living things.

Compelling grounds can be provided for many of these responses to the agency question, and passages from Plotinus' *Enneads* can be found that would seem to advocate them as well. In *Enn.* 5.7.2 Plotinus accounts for the various features of the offspring in terms of the mixture of the *logoi* present in the parents' (lower) souls such that different features result depending on which parental *logos* of a given feature ends up dominating the mixture.[40] This approach has the obvious advantage of providing a direct explanation for the offspring's resemblance to its parents. But in *Enn.* 6.7.7.8-16 and 1.1.11.8-15 Plotinus seems to associate the task of generating an individual body first with the World-Soul, which is said to provide a sketch – a sketch that is then articulated and made complete by the offspring's own individual soul, without so much as mentioning the parents.[41] This approach seems to gain some support from the Myth of Er (*Republic* 617D-620D) where individual souls are said to be themselves responsible for their future lives, and perhaps by *Phaedrus* 248D and *Laws* 967D where souls are arguably associated with seeds.[42] The World-Soul's own role seems even to go beyond the mere providing of a sketch, when one focuses on two views that Plotinus often advances: first, that the (higher) individual soul only descends to a body that is already suited to it,[43] and second, that all lower soul in some sense belongs to the World-Soul.[44] The advantage of making the World-Soul the primary agent of human embryo formation is that it would seem to be the only viable candidate for the soul responsible for generating would-be 'spontaneously' generated creatures such as worms and gnats, and so for reasons of economy as well as in order to avoid the potential blasphemy of claiming that the noble World-Soul is responsible only for the most ignoble creatures, one might extend the scope of its responsibility to include humans and all other living creatures.[45] As discouraging as Plotinus' various remarks might be to one seeking a single, coherent embryological aetiology, these apparent discrepancies need not be irresoluble, but they do go to show both how tricky the metaphysics of embryology can be for a Neoplatonist and how fortunate we are to possess Porphyry's deliberations on this topic.

Porphyry's own views seem to be as follows.[46] The father's lowest (*phutikê*) soul power is the sole producer of seed.[47] Regarding the manner of production, Porphyry's stance on whether this is a case of generation (*gennan*) or of splitting-off (*apomerizein*) seems at times somewhat non-committal.[48] This is significant because basic principles of Neoplatonic metaphysics would require that any product of generation be itself inferior to (insofar as it is a mere image of) its generator,[49] whereas what is separated off would supposedly be on a par. Thus, we can see some oscillation in Porphyry's description of the seed itself. Sometimes he suggests that it has its own *phusis* from the father,[50] while at other times he insists that the seed, being the product of *phusis*, must itself be something inferior insofar as a *phusis* possesses actual motion while the seed itself does not.[51] This tension can perhaps be eliminated by distinguishing between engaged and disengaged natures. Only engaged natures possess actual motion, and being engaged is a matter of being linked into the ontological hierarchy. The father's *phusis* is engaged because it is mixed with his irrational soul (which in turn is connected to his rational soul, and so on), and the nature in the seed, too, is engaged as long as it in a similar position. But when it is cut off from this chain, the *phusis* in the seed is no longer engaged. Hence, the picture of the seed that we have here is very un-Aristotelian. It is not an entity that contains its own principle of motion (unlike Aristotle, Platonists make soul rather than mere nature the *arkhê* of motion and change) that only needs matter to get going. The Neoplatonic seed needs help from both sides, as it were. It needs to be engaged by the next soul up in the ontological hierarchy in addition to being supplied suitable matter.

This *phusis* in the seed undergoes three stages. At its inception the *phusis* in the seed is still blended with and thus engaged by the father's own *phusis*. After perhaps a moment of disengagement, the *phusis* in the seed combines with the mother's nature at conception, thus being re-engaged. Finally, at birth the *phusis* is disengaged from the mother's soul in order to blend with its own descended soul. In describing these three stages Porphyry explicitly appropriates the Stoic doctrine of complete blending, which he takes to be incoherent in its inherited form, as bodies cannot completely interpenetrate each other, but a rather fitting account of how different soul-powers can form a unity.[52] This appropriation is particularly important because of what it tells us about the *phuseis* of the parents and offspring. The crucial feature of Stoic blending – what distinguishes it from mere Aristotelian mixtures – is that each blended component retains its own identity in such a way that both components can again be separated from each other.[53] Hence this application of total blending to the parents' and offspring's *phuseis* makes sense only if these *phuseis* are in some important respects distinct from each other. In other words, it does not seem right to say that soul at the level of *phusis* is not individualized in any way, that there is simply a single universal *phusis* at work in the world.[54] Rather, each individual has a nature suited to it that accounts for the various features of its body and that differs significantly from that of each of its parents.

Originally the offspring inherits its *phusis* from its father. This *phusis* is in fact a replica of its father's own *phusis*, which is to say that all the *logoi* in the father's *phusis* are found again in the offspring's (at stage one). But at conception, when the seed's *phusis* blends with the mother's, the offspring's *phusis* undergoes certain changes. This is presumably because the potential *logoi* in the seed's disengaged *phusis* themselves need to be actualized by actual *logoi* of the same kind.[55] For example, the seed's *logos* for nose, inherited from the father, is potential just prior to conception, and needs to be actualized by the *logos* for nose in the mother, which is actual since her *phusis* is, unlike the seed's at this point, anchored in a higher substance, namely her own irrational soul. The engaging of the seminal *phusis* is simply a matter of having all its *logoi* actualized in this manner. This engagement can lead to changes in the seminal *phusis* when the actual corresponding *logoi* in the mother's *phusis* are of a different type. If the seminal *logos* of nose (from the father) is for a snub-nose and the actualizing maternal *logos* of nose (corresponding to her own physical features) is aquiline, the resulting actualized seminal *logos* of nose might either remain snub-nosed or become aquiline.[56] These same mechanics of actualization serve to explain another phenomenon of change, namely how the offspring's appearance can be affected by the mother's perceptions at the time of conception.[57] Though this no doubt strikes us as odd, there was a wide consensus in antiquity – even among doctors – that such a thing happens,[58] and so it is a desideratum of any metaphysical account of embryology to explain how this phenomenon of ideoplasty can take place. Porphyry's embryology of engagement and actualization can do precisely this. The idea seems to be that the conception is a moment of metaphysical reversion to higher principles. Under normal conditions the potential *logoi* in the disengaged seminal *phusis* revert to and are actualized by the corresponding actual *logoi* in the mother's nature, but it seems possible for them to bypass these *logoi* and revert, as it were, to greater heights, thus being actualized by and perhaps even assimilated to the *logoi* in her *phantasia*. Thus, two dark-skinned parents, having *logoi* for dark skin in their *phuseis*, could produce a light-skinned child, if the mother has acquired a *logos* of light-coloured skin by looking at (even an image of) a light-skinned individual at the moment of conception.[59]

This is how the mother is able to contribute to the physical appearance of the offspring without supplying any seed. Throughout the period of gestation the mother continues to occupy the role of actualizer with respect to the offspring and thus presumably continues to influence the offspring's physical constitution, yet it would seem that offspring's *phusis* has been for the most part determined at conception, since the *logoi* in the mother's *phantasia* seem at least to have no effect on the offspring after that. Hence, Porphyry replaces Aristotle's characterization of the embryo's relation to the mother as analogous to that of a plant's relation to earth with a grafting metaphor which grants the mother a much more active role in the formation of the offspring. Unlike Theophrastus who did not detect any meaningful difference between, on the one hand, the relation of the grafted scion to the rootstalk and, on the other hand, the relation of a plant to the earth – both the rootstalk and the earth simply served to provide nourishment,[60] Porphyry sees grafting alone as illustrating the kind of dual input briefly discussed above with respect to the seed. The rootstalk not only supplies nourishment to the scion, but also serves as its source of psychic power as it were, which is to say that it engages the *phusis* in the scion and actualizes its *logoi*. As a consequence of this, the fruit and flowers of grafted scions can take on some of the properties of their rootstalks. Porphyry must have been aware of such cases of grafting chimeras (as presumably Theophrastus was not), and used this as grounds for his metaphysical speculations on the mother's seedless influence over the offspring.[61] This new understanding of the mother's role in the formation of the offspring emboldens him to refer to the mother as a *demiurge*[62] – an accolade that Aristotle had reserved exclusively for the father.[63]

Thus the chief agent in the production of the offspring's body is its own *phusis*.[64] The father's *phusis* has the important task of initially providing this *phusis*, while the mother's soul, which Porphyry categorizes as a co-cause (*sunergon*),[65] is actively involved in engaging the offspring's *phusis* and using her position as captain[66] to steer the construction in certain ways, and in this not only her *phusis* but even her *phantasia* is involved. The offspring's own descended soul has no role to play in the actual formation of the embryo, nor do the rest of the parents' souls, except insofar as they serve to support these activities. This leaves the question regarding the influence of other external factors, in particular matter, the environment and the World-Soul. Porphyry does not really give an account of teratology in *Ad Gaurum*,[67] but presumably he would be in agreement with Plotinus[68] that deformities and deviations from nature's plan are largely due to matter.[69] Environmental factors could include any number of things, e.g. the material make-up of the seed,[70] nourishment and the uterine environment,[71] as well as geographical[72] and astrological[73] factors. There appear to be only two statements on the environment's influence in the *Ad Gaurum*. At one point Porphyry rather opaquely tells us that the environment (*to periekhon*) also supplies movement to the offspring's *phusis*, in a manner similar to that in which the mother's *phusis* is supposed to supply movement.[74] This is a somewhat troublesome remark, as he repeatedly says that the mother is herself sufficient as an actualizer of the potential motion in the offspring's *phusis* for as long as it is in gestation,[75] so there should be no need for the environment to act as a source of motion at this time. By contrast, later on, Porphyry seems to distinguish the mother's contribution from that of the universe (*to pan*) both in terms of the time periods in which each is efficacious and in terms of the objects of their respective concerns.[76] Whereas the mother's activity is directed at the offspring's *phusis* during gestation, the universe acts on 'the irrational *corporeal* substance' after birth.[77] There is no mention here of anything other than the mother assisting in any capacity in the development of the offspring before birth, nor of the universe or environment acting as a source of motion to the offspring's *phusis* at any time (after birth it is in the care of its own descended individual soul). Perhaps these two passages can be harmonized in the following way. It would be hard to deny that the nature of the external environment during gestation has some effect on the physical constitution of the offspring. Insofar as all of our bodies are ultimately constituted of the four elements, elemental factors such as heat and moisture surely have some effect on the formation of one's individual physical features, and these elemental factors themselves result from *logoi* which as enmattered form-principles fall under the domain of nature. Thus the external environment, by providing such things as heat and moisture, should have some effect on the offspring's physical constitution and even on its *phusis*, but this hardly compares to the influence that the mother herself exercises over it. After birth the external environment's influence on the body grows considerably and this again might have some marginal effect on the offspring's *phusis*, but it is primarily the descended individual soul that accounts for its actual motion.[78] The World-Soul does not seem to figure prominently into *Ad Gaurum*'s embryology. Its primary task seems to be that of a divine match-maker – ensuring that descending individual souls find their way to bodies suited perfectly to them.[79] But of course to whatever extent the environment or the changing seasons affects the development of the offspring, we might say that the World-Soul is at work.

3. The text

This translation of Porphyry's *Ad Gaurum* is based on Kalbfleisch's edition of the text printed in the *Anhang zu den Abhandlungen der königlichen preussischen Akademie der Wissenschaften* (Berlin, 1895).[80] This text is based on a single twelfth-century manuscript (Cod. Suppl. gr. 635, Paris, Bibliothèque Nationale),[81] which is extremely valuable in that it is our only source not only of the *Ad Gaurum* but also of Galen's *Introduction to Logic*, yet it is also, regrettably, not entirely reliable. Preceding the 21 folios containing these two treatises were at least 88 folios[82] of other material, including Galen's *On Marasmus*, only the conclusion of which is preserved. A comparison of this conclusion with the *vulgata* reveals a number of errors, and this negative assessment of the scribe's diligence has been corroborated by the extent of editorial revision required by the *Introduction to Logic*.[83] To make matters worse, the manuscript contains many abbreviations, not only tachygraphic ones with established meanings but also less conventional abbreviations that can be understood only in context. It has also suffered some water damage on the top and outer edges, thus making some passages very difficult to decipher; indeed, the last few pages are half-destroyed. Hence some restoration and conjecture were required. Prior to Kalbfleisch only Minoides Mynas, who discovered the manuscript in the monasteries on Mount Athos,[84] had himself produced a copy of the *Ad Gaurum* (now Paris Supp. gr. 727), which included several textual changes, few of them, in Kalbfleisch's opinion, hitting the mark. Thus, in the critical apparatus he includes only those of Mynas' proposals that he deems 'either correct or at least noteworthy'.[85] For this reason I have not consulted the copy by Mynas. Kalbfleisch's own extensive textual emendations and his suggestions in the critical apparatus and *Nachträge* (p. 80) are extremely helpful, as are those proposed by Festugière in the notes to his translation. In this translation I have only rarely departed from Kalbfleisch's text (see *Textual Emendations and Conjectures*), and in every instance I provide my reasons for doing so in an accompanying note. Very often I depart only to follow some suggestion that Kalbfleisch himself has made in the apparatus or *Nachträge*, or else one by Festugière, though in a handful of cases I deviate from both.

Kalbfleisch uses parentheses, ( ), to mark places where he has been able to reconstruct the text where it has been destroyed. There are uncountably many of these, and in most cases they concern individual letters rather than entire words, making it unclear how to mark them in the translation. Even if that were clear, to have marked every such case in the translation would have made for very tedious reading. To give just one example: *AG* 2.5.3-4 would end up looking something like this: '(And even if) we (we)re to conc(ede) that (th)e embryo itself i(s) potentially or even actually an animal, we (will still maintain) ...'. Thus, I have signalled these restorations – using curly brackets, { }, in order to avoid syntactical misunderstandings – in the translation only in a few rare cases where I disagree with Kalbfleisch's suggestion (e.g. 17.2) as well as in the notes at times in order to convey how uncertain a given passage is. Destroyed text that Kalbfleisch has not been able to restore is marked by a sequence of dots, ..., corresponding to the probable number of letters. These I have generally reproduced in the translation (e.g. 17.4). In some cases I have filled these gaps in the translation, either on my own or following Kalbfleisch's notes or Festugière's translation. In such cases I replace the periods with the suggested translation in curly brackets thereby indicating that it is a restoration (e.g. 16.7.4-5). I do not indicate any of the insertions, typically marked with angle brackets, < >, that Kalbfleisch himself made in the restoration of the text, but I do use angle brackets to indicate any insertions not already included in the body of Kalbfleisch's printed text (e.g. 11.2.13). I use square brackets, [ ], to indicate content that is only implicitly contained in the text but that is desirable to make more explicit in the translation. Here again there is the danger of cluttering the text and making it unreadable, and so I try to use these sparingly.

I have retained Kalbfleisch's division of the text into chapters and sections, though there are no such divisions in the MS. These are important above all because the *TLG* organizes the text according to these chapter and section numbers, rather than by the page and line numbers of Kalbfleisch's text. In the introduction and notes I follow the *TLG*'s practice and refer to the text by chapter, section and line, with the exception of the last chapter, which Kalbfleish was unable to organize into sections on account of its incomplete state.[86] When referring to a passage from this chapter, I give the chapter number together with the page and line numbers of Kalbfleisch's text (e.g. 18, 61,6-7 K).

Currently there is work being done by a team of scholars led by Luc Brisson on a new text and French translation.

4. Issues of translation[87]

In the language of modern medical texts it is customary to distinguish between three stages in the development of the offspring: the blastocyst (roughly first two weeks of pregnancy), the embryo (third to ninth weeks), and the fetus (from the ninth week onwards).[88] One can also find some distinctions between various stages of gestation in Greek embryological literature. In *On the Seed*, for example, Galen puts forward a four-stage development that is reflected in a stage-specific terminology for the offspring: (1) *gonê*; (2) *kuêma*; (3) *embruon*; (4) *paidion*. While the form of seed is predominant in the first stage, the second phase is characterized by the presence of blood as well as the liver, heart and brain in a very primitive form; in the third phase the liver, heart and brain are now formed and identifiable while the other parts are there in outline; in the fourth (still prenatal) stage, the offspring's limbs are completely differentiated and it 'moves just like a completed animal'.[89] Yet despite Galen's claims regarding the Hippocratic origin of this division and vocabulary,[90] this division and its vocabulary are far from being universally accepted. Indeed, Galen himself proposes a very different schema of four stages in his treatise *On the Formation of Fetuses*: In the first stage arteries, veins, the chorion and liver are constructed; in the second stage the heart is built; in the third stage the brain and facial features are formed; in the fourth phase, which in this treatise now begins after birth, teeth and hair grow while the other parts are strengthened.[91] This lack of universally accepted stages of embryonic development is reflected in the language of Greek embryology, and this allows Porphyry here to use *to embruon* to refer to every stage of the offspring's development right up to just prior to the moment of delivery.[92] I have decided to translate *to embruon* consistently throughout the text as 'embryo'. This results in a few slightly jarring sentences, but this seems preferable to the alternative of artificially trying to impose developmental stages onto the text. Indeed, it might be part of Porphyry's rhetorical strategy for establishing the vegetable status of the offspring in the womb to refer to it even in the late stages of gestation as an *embruon*.

A similar problem is encountered with the passive participle *to kuoumenon*. The participles *hê kouousa* and *to kuoumenon* are commonly used to describe the pregnant mother and unborn offspring respectively. The primary sense of the active verb *kuein* (which appears much less frequently than the participles)[93] is 'to bear in the womb' or 'to be pregnant' so that the active participle can simply be translated as 'pregnant'. Similarly, the most straightforward translation of *to kuoumenon* would be something along the lines of 'the thing carried in the womb'. This is a sense that obviously applies to the offspring throughout the entire pregnancy (after birth, Porphyry uses the aorist '*kuêthen*' to refer to the offspring). Nevertheless, 'the thing carried in the womb' is too unwieldy to be serviceable in a translation, especially given the frequency with which it appears in the text. I have therefore decided to translate it with the traditional 'fetus', partially in the hope that this might balance out the use of 'embryo'.[94] Here again, since 'fetus' has a narrower meaning today, the reader must bear in mind that 'fetus' and 'embryo' in this translation are being used synonymously to refer to the unborn offspring at every stage of the pregnancy. There are, however, two cases where the unwieldy translation seems preferable. The first is when Porphyry uses *kuoumenon* with another noun such as *sôma*: 'the body that is being carried in the womb' (cf. 12.1.3). The second concerns one of Porphyry's arguments (7.3) that seems to me to depend on this sense of being carried in the womb.

One of Porphyry's main concerns in *Ad Gaurum* is to establish that the embryo in gestation is analogous to a *phuton* rather than a *zôion*. Hence, although *zôion* can have a broader sense of 'living thing' that extends to include even plants, here it must have the narrower sense of 'animal', and I translate it thus throughout. For the more inclusive sense Porphyry uses the participle *zôn*.[95]

Porphyry's expression *hê phutikê* (6.3.2; 6.4.1; 10.5.2-4; 14.3.1 and 9; 15.5.1) can either be taken to be referring to the vegetative *power* of soul (*hê phutikê dunamis*) or simply to a vegetative soul (*hê phutikê psukhê*).[96] Porphyry would allow either of these possibilities,[97] though he seems to prefer 'power' and so I have generally supplied this in the translation. We can witness this flexibility and preference in 16.1.1-6: 'Further, let the seed have a soul – if someone wishes to call the vegetable power a soul [...] for even Hippocrates seems to call the vegetable power a "soul" in accordance with ancient usage'. From this we can see both that he prefers to refer to the *phutikê* as a power, and that he is willing to call it a soul. Both of these attitudes can also be witnessed elsewhere. Porphyry does at times supply *dunamis* himself (9.2.7; 17.3.4), and in 10.5.4-5 he is clearly opposing the mother's *phutikê* with her soul. But he also supplies *psukhê* at least once.[98]

Another difficulty is presented by *hê gastêr*. In the Hippocratic corpus it is often used more or less synonymously with *mêtra* (or often the plural *mêtrai*)[99] to indicate the womb,[100] whereas Galen to the best of my knowledge reserves *hê gastêr* to use more narrowly of the stomach.[101] Porphyry follows the Hippocratic tradition in this regard, and he uses both *hê gastêr* and *he mêtra* regularly. However, he also uses *hê gastêr* in a digestive context to refer to the stomach (7.2) so that a more inclusive translation such as 'belly' seems desirable. In order to capture the synonymy of *hê gastêr* and *he mêtra* without eliminating the linguistic distinction, I have chosen to translate the former with 'belly' and the latter with 'womb'.[102]

All remaining translational issues are discussed in the notes.

5. The content of the *Ad Gaurum*

Porphyry signals that the *Ad Gaurum* is divided roughly into four sections:[103] (A) an introduction (1.1-2.5), (B) arguments aimed at showing that the embryo or fetus is not actually an animal (3.1-12.7), (C) arguments aimed at showing that the embryo is not even potentially an animal (13.1-16.9), and (D) arguments aimed at showing that the soul necessarily comes into the offspring from outside (i.e. not from the parents), even if it is conceded, *per impossibile*, that the embryo or fetus already has a soul (17.1-18 end). I give a brief summary of the contents below, focusing mainly on the arguments, objections and replies that Porphyry presents in favour of his views.

(A) *Introduction*. The main question concerns the psychological status of human embryos (or fetuses): Are they already in possession of the self-moving soul and thus already animals? Porphyry distinguishes three possible positions: an embryo is actually an animal, it is potentially an animal in the sense that it possesses the soul but does not use it (henceforth 'potentially$_2$'), or it is potentially an animal merely in the sense that it can eventually receive a soul ('potentially$_1$'). As the defining psychological features of animals are their powers of sensation, impulse and movement, the following examination focuses on these powers [1.1-4]. One major advantage to the view that it is merely potentially$_1$ an animal, is that this establishes a clear moment at which the soul enters, namely at birth. By contrast, those who insist that it is actually or potentially$_2$ an animal disagree about when the soul enters. And there will be an analogous issue concerning Plato's views on the moment of entry [2.1-4].

(B) *Arguments aimed at showing that the embryo is not actually an animal.*

Argument B1: The nutritive administration of embryos is like that of plants, not animals. (a) Plants nourish themselves through the roots while animals nourish themselves through the mouth (processing it through organs) [3.2]. Embryos are not nourished through the mouth [3.4] (b) Animals breath through their noses and perish when immersed in a moist environment, but plants breath through 'heartwood' and require a moist environment [3.3]. Embryos do not breathe through the nose and require a moist environment [3.4]. Thus, the animal soul is not actually present. (cf. Argument 3 below).

(Platonic) Objection: This turns on the ambiguous relation between the Platonic division of soul and the Aristotelian one. Aristotle divides soul into various faculties responsible for nourishment, sensation, desire (impulse), local motion and thought (e.g. *De Anima* 414a31-2), and defines animals by their possession of the powers of sensation and desire (414b1-16). But it is unclear how to map this Aristotelian division onto the Platonic division of appetitive, spirited and rational parts, and in particular how to describe the lowest kind of Aristotelian soul – the vegetative soul responsible for nourishment and generation – in Platonic terms. Porphyry's Platonic objectors capitalize on this ambiguity in the following way: Plato describes the vegetative soul in the seed as part of the lowest part of soul – appetitive – and he attributes sensations of pleasure and pain to it as well as impulses in the form of desires. Since these are the defining Aristotelian characteristics of animals, seeds (and thus embryos) must be animals [4.1]. *Reply*: This understanding of the vegetative soul effectively eliminates the distinction between animals and plants [4.2]. Plato calls plants and seeds 'animals' (*zôia*) merely in the sense of being alive (*zên*), but not in the more robust sense of having a self-moving soul [4.3-5]. One must understand Plato's claim about plants (and thus embryos) having *aisthêsis* in light of the distinction between sensation and perception (though Plato uses the term 'opinion' instead of 'perception'). Sensation refers to the passive and more or less physical process that goes on in sense-organs, while perception refers to the actual phenomenal experience of, e.g. seeing a colour. Plants, and thus embryos, merely undergo these passive physical motions, but they do not perceive, and so they do not actively have the self-moving soul that is responsible for perception and that makes something a genuine animal [4.6-11].

(Empirical) Objection: (a) There seem to be empirical grounds for

attributing impulse and local motion to fetuses. For movements can be perceived when they are still in the womb, and they seem to be actively involved in their own deliveries. (b) There is also evidence for attributing the faculty of desire to them, for the pregnant mother's cravings would seem to be due to the fetus [5.1-3] (c) Finally, the phenomenon of ideoplasty would seem to indicate that they must even have a faculty of representation [5.4-5]. *Reply*: (c) The embryo need not have a faculty of representation in order to explain ideoplasty. Ideoplasty occurs when the embryo's vegetative soul blends together with the mother's vegetative soul and in this way becomes obedient to her faculty of representation [6.1-4]. (a) Nor does it follow from the movements undergone by the fetus that it must have faculties of representation and voluntary motion, since there are many examples of bodily motions that are not accompanied by either [7.1-3]. (b) The pregnant mother's cravings do not originate in the fetus but in the womb. For Plato describes the womb as an animal with appetites, impulse and even motion [8.1-4].

(Platonic) Objection: In the *Timaeus* Plato describes conception as the planting of 'animals' in the womb that are too small to see. Thus, Plato is a preformationist who thinks that embryos are animals from the very start [9.1-2]. Moreover, this is corroborated in the *Phaedrus* and elsewhere where he seems to say that individual souls go into the seed or enter at conception [9.3-4]. *Reply*: Plato is not a preformationist. In the *Timaeus* he emphasizes that these so-called 'animals' are in fact 'unformed' and what does not have a form cannot be an animal yet [9.2]. In the *Phaedrus* one must not understand *gonê* to be referring to seed but more generally to generation, and likewise in the other passage Plato is merely saying that the soul comes to belong to the class of things that are conceived, that is to say, to mortal bodies [9.2-4]. The *Timaeus* and *Phaedrus* make clear enough that the soul enters only once the body has been completed [9.5].

Porphyry's own embryological account: The process of the formation and development of the embryo would appear to be most like the process of the formation and development of a fruit from a scion grafted onto a rootstock. The salient features of the grafting process that are found again in embryology are: (a) The physical development of this scion's fruit also mirrors that of the embryo for the most part: The seed/scion constructs first a membrane/pod, then an umbilical cord/stalk from which the embryo/fruit can draw breath and nourishment, finally the interior parts are formed and made solid [10.2-3]. (b) More importantly, the scion does not just receive nourishment from the rootstock, rather its nature forms a natural unity with the rootstock's nature with the result that the scion's fruit can inherit some of the rootstock's physical characteristics [10.1-2]. In embryology the nature in the seed forms a natural unity with the mother's nature and can thus inherit some of her physical characteristics. In each case the nature of the thing coming to be needs to be ontologically grounded in a higher soul principle – in the one case that of the rootstalk (which is then in turn grounded in the soul of the earth)[104] and in the other case a self-moving soul [10.1 and 10.4-5]. At birth the individual self-moving soul arrives instantaneously – not because it is compelled – but on account of the suitability that obtains between individual bodies and souls [10.6-11.2, cf. 13.7]. This suitability is, as it were, managed by the World-Soul [10.6 and 11.4].

Argument B2: The entrance of the soul is similar to the entrance of the intellect. Everyone agrees that the intellect's activity starts well after birth, just as the soul's starts well after conception. In each case it is critical that the underlying subject first be made suitable to the activity [12.1-3].

Argument B3: Although both are incomplete, there are essential differences between fetuses and newborns. (a) Newborns eat real food through their mouths as opposed to receiving other nourishment through an umbilical cord [12.4]; (b) newborns use crying and gestures to signal what causes them pain, while fetuses are completely silent and without representation [12.4]; (c) rational part of soul is potentially present to newborns insofar as they have sensation, whereas fetuses do not even have this [12.5]; (d) newborns move while the fetuses do not – all would-be motions of the fetus are in fact due to the mother's body [12.6]. Thus, these soul-powers are actually present in the infants but not in the fetuses. Cf. Argument B1 above.

Argument B4: The powers of soul characteristic of animals cannot be active in the embryos, since the activity of these soul-powers would not be merely superfluous, but would actually hinder the activity of the generative and nutritive powers [12.7].

(C) *Arguments aimed at showing that the embryo is not potentially$_2$ an animal.*

Argument C1: A form or power is potentially$_2$ in an underlying substance only when that substance has been suitably organized. As long as it is not suitably organized, the form or power can only be said to be in it potentially$_1$ [13.1-3]. The underlying substance of soul-powers are the organs, and so soul-powers can only be potentially$_2$ present once the organs have been completed. This is at birth, as completion in nature is accompanied by departure [13.4-5]. This goes for Plato, too, since he locates the three parts of soul in organs, which are again not complete until parturition [13.6]. Cf. Argument C5 below.

Objection: Soul-powers are just like everything else contained potentially$_2$ in a seed – animals possess them from the beginning in the form of *logoi*, long before these things reveal themselves. Teeth, for example, only appear later, though the offspring had the *logoi* for teeth all along. Thus, the soul-powers are also potentially$_2$ present even in the fetus [14.1]. *Reply*: Souls cannot come to be from a seminal state. For that would imply that an ontologically lower nature is capable of producing an ontologically higher nature, which is impossible, since the product is always inferior to the producer [14.2-4].

Argument C2: With the self-moving soul there is no such thing as a purely potential$_2$ state. Even in sleep it is slightly active through representations in dreams [15.2].

Objection: The completed organs are not necessary for sensation because the soul is even able to perceive in a disembodied state, according to Plato and many others. *Reply*: It is necessary to distinguish between the soul's perceiving and the composite living thing's perceiving. Only the latter qualifies the living thing for animal status, and this requires completed organs [15.3-4].

Objection: The organs need not be complete. For nature is also a power of soul, and it can be active even when the body that it is present in has all sorts of deficiencies. Thus, this should be possible with other soul-powers too. *Reply*: There is empirical evidence that shows that the other powers of soul are inhibited by physical deficiencies in their proper organs in a way that nature is not [15.5].

Argument C3: The rest of the soul cannot be present to the embryo [even potentially$_2$] because it would be there in vain and even obstructive, since the vegetative power and the other powers inhibit each other. Whatever external support the embryo's nature requires, it receives from the mother – anything more would be an obstruction [16.1-4]. After birth it receives further support for its continued development from its own individual soul [16.5].

Argument C4: Astrologers also tend to think that the soul enters at birth [16.5-6].

Argument C5: It follows from the harmony/suitability doctrine, that the soul is not even potentially$_2$ present if the matter is not suitable, just as the harmony is not potentially$_2$ present in an untuned instrument nor sight in a damaged eye. Since the soul's use of the body is an all or nothing affair, it cannot be present to a body that is not completely ready for it [16.6-9]. Cf. Argument C1 above.

(D) *Arguments aimed at showing that the soul comes from outside (i.e. not from the parents), regardless of whether this is believed to happen before birth or not.*

Objections to external source of soul: (a) It is absurd to say that the parents do not produce the soul but only the body, because this would entail that man does not produce man, even though horses produce horses and wheat wheat [17.3]. (b) It is only in cases of spontaneous generation that the substantive form comes from outside, but human generation is not spontaneous [17.4-5]. These problems can be solved by simply saying that the powers of soul are potentially there from the start, inherited from the parents, and become actualized later on just as other, physical features do [17.6-7]. Moreover, the seed is the immediate product of the vegetative soul but it is transitively the product of the entire ontological chain of souls, so it is reasonable for these powers to be contained in the seed in a potential manner. This is corroborated by the close connection that obtains between the power of representation and sexual desire [18, 60,25-61,13 K]. *Reply*: The reply must remain speculative for the most part, as our text becomes largely unreadable at this point. Porphyry's reply seems to begin by attempting to determine and address the motivations one might have for denying that the soul comes in from outside. There appear to be three possible motivations (these do not correspond to the objections (a) and (b) above): (i) Their reason for putting this view forward might be that they are worried that the external soul thesis would lead to an infinite number of souls, but this is not so on account of transmigration [18, 61,22-62,2 K]. (ii) Perhaps they advance this view because they are impressed by the character similarities between parents and their offspring, and so conclude that the offspring are inheriting soul as well as body. But character similarities can be explained by the external thesis, too. The parents create a body that is similar to their own bodies, and then by the doctrine of suitability a soul will enter that is suited to this body and thus shares a certain character with the parents [18, 62,2-9 K]. (iii) These men oppose the external thesis because they say human generation ceases to be a case of like generating like and becomes more like spontaneous generation where unlike generates unlike. But their internal view falls victim to the same problem. For they want to say that the vegetative soul produces the rational, self-moving soul, where the latter is immortal and knows what it does while the formal is mortal and acts without knowledge [18, 62,9-30 K].

Notes

1. *de Mar.* (Kühn 7.666-704).

2. *Inst. Log.* (not in Kühn). For a list of editions and translations of this and all other Galenic and Pseudo-Galenic texts, see Fichtner (1997) and Hankinson (2008) 399-403.

3. He also suggests (25n.1) that Porphyry's given name (Malchus), which is the name found, for example, on his *Life of Pythagoras*, could conceivably have been corrupted into Galen.

4. The authoritative edition is Nickel (2001). The text is also part of Kühn's edition (4.652-702). An English translation can be found in Singer (1997) 177-201.

5. The authoritative edition is De Lacy (1992), which also contains an English translation. The text is also part of Kühn's edition (4.512-651).

6. The authoritative edition is Helmreich (1907-9). The text is also part of Kühn's edition (3.1-4.366). An English translation is offered by May (1968).

7. *de Foet. Form.* 74,11-16 Nickel = 4.670,12ff. Kühn.

8. 2.357,24-8 Helmreich = 4.329,1-4 Kühn.

9. Unfortunately, the relevant parts of *de Demonst.* and *de Plac. Hipp. et Plat.* have apparently not been preserved. May (1968) reasonably suggests (668n.26) that this might have been contained in the first four to five chapters of Book 1, which have not survived, though De Lacy reconstructs some of the content of these chapters from preserved testimony and fragments (see De Lacy (1984) 64-71). Yet May does point out that this belief is implied at several points throughout *de Plac. Hipp. et Plat.* (Book 2, section 8; Book 5, section 6; Book 6, section 6).

10. *AG* 13.4; cf. *de Foet. Form.* 78,8-9 Nickel (= 4.674 Kühn).

11. *AG* 13.6.

12. As Galen himself admits at *de Foet. Form.* 66,19-32 Nickel (= 4.663-4 Kühn) and *de Prop. Plac.* 90,17ff. Nutton. Note that Nutton's edition must now be checked against the more recently discovered Greek which was edited by Boudon-Millot and Pietrobelli in Boudon-Millot and Pietrobelli (2005) and also available in Boudon-Millot (2007). See also Nickel (1989) 80-2.

13. cf. *de Foet. Form.* 72,8 Nickel (= 4.668 Kühn).

14. *de Foet. Form.* 68,9ff. Nickel (= 4.665 Kühn).

15. *de Foet. Form.* 70,12-13 Nickel (= 4.667 Kühn).

16. *de Foet. Form.* 74,11-18 Nickel (= 4.670-1 Kühn).

17. See G.E.R Lloyd, 'Fuzzy Natures' in his (1996).

18. *de Foet. Form.* 76,10-78,11 Nickel (= 4.672-4 Kühn).

19. cf. Iamblichus *de An.* (Finamore and Dillon) §31.

20. 2.357,10-21 Helmreich (= 4.238,4ff. Kühn). Note that M.T. May gets this passage wrong in her translation of *de Usu Part.*, as she incorrectly suggests that Galen did *not* think fetuses moved by impulse. In this passage, as I understand it, Galen is censuring the physicians in question for not being aware of three facts: (i) that the natural (i.e. *not* voluntary) position of the muscle situated at the urinary channel is to keep the ureter closed, (ii) that fetuses already make use of voluntary activities involving impulse, and (iii) that when the completed animal chooses to urinate, its voluntary motion involves relaxing the naturally closed position of that muscle along with contracting the bladder around the liquid. That the physicians simply denied (ii) is clear from 2.357,4-5 Helmreich. (i) and (iii) appear to register Galen's dissatisfaction with the physicians' general statement about the muscle situated at the urinary channel 'serving' (*hupêrtôn*) by means of voluntary action (2.357,7-10). In fact, it serves to keep the ureter closed without voluntary action, while the voluntary actions involved in urination are more numerous than their reasoning allows. For even when we grant the physicians that the fetus' urine passes through the umbilical cord and not the ureter, it would still appear that voluntary action might be needed – namely contracting the bladder. So their argument that the absence of voluntary action implies that urination must proceed through the umbilical cord fails. I would like to thank Philip van der Eijk for some helpful suggestions regarding this passage. Cf. *de Mus. Dissect.* 181,10-14 Debru and Garofalo (18B.998 Kühn); *de Usu Part.* 5.14 and 16.

21. *in Hipp. Epid. 6 comment.*, 17A.811,12ff. Kühn. Cf. *de Plac. Hipp. et Plat.* 402,32-7 (= 5.558 Kühn)

22. *de Foet. Form.* 102,27ff., esp. 104,12-26 (= 4.699,3ff. esp. 700,2-6 Kühn). See also *de Plac. Hipp. et Plat.* 596,5ff. De Lacy (= 5.809 f. Kühn), and *de Prop. Plac.* 90,17ff. Nutton (though see note 12 regarding this text).

23. See Introduction §2.

24. See Lesky (1951) 177-93 and Nickel (1989) 29-49.

25. See Introduction §2.

26. See Wilberding (2008) and Introduction §2.

27. For a recent discussion of Galen's embryology, see Boudon-Millot (2008).

28. Kalbfleisch picks out several terms for which one can find many Neoplatonic parallels: *hupobainein* (*hupobasis*), *epistrephein* (or *epistrophê*), *athrous* and *en diexodôi*, *thixis* (and *epaphê*), and *hê alogia* for the lower parts of the soul (see Kalbfleisch 14n.2).

29. One might add that Iamblichus is also eliminated for doctrinal reasons, see above note 19.

30. See the note on *AG* 6.1 below.

31. Dörrie (*Summ. Zêtêm.* 161-5) takes issue with Kalbfleish's claim that Porphyry would refer to the *Summ. Zêtêm.* as an *hieros logos*, but does agree that the *AG* is by Porphyry. See note *ad* 10.6.1-3.

32. Psellus concludes his report, after remarking that the embryos are nourished through the umbilical cord and not through the mouth, with 'hence, the midwives close off the umbilical cord so that the offspring might be nourished through the mouth'. Nothing in *AG* corresponds to this last suggestion that the umbilical cord closure is somehow connected to this switch to nourishment through the mouth. Kalbfleisch (25) suggests this is either Psellus' own addition, as it is in direct speech, in contrast to the preceding, or else contained in the lost conclusion of *AG*.

33. *Enn.* 4.7.8^{3}.9-11; 3.8.5.24-5; 5.1.6.37-9; 5.1.7.47-8; 5.5.13.37-8; 5.8.1.19-21; 6.7.17.4-6.

34. *de Myst.* 3.20.

35. *Elem. Theol.* §7.

36. e.g. *Enn.* 1.9.1.4-7; 4.3.8.50-4; 4.3.12.37-9; 4.3.13.7-12; 6.4.15 *passim*. Regarding (4) cf. *Enn.* 1.2.4.25-9 and 4.4.2 *passim*; Iamblichus *Protr.* 21.

37. cf. Plotinus *Enn.* 4.5 *passim* (cf. 1.6.9.29-30) and Iamblichus *de Comm. Math.* §8.

38. A case in point is the *de Abst.* Clark (2000) 7-8 discusses the absence of any mention of Plotinus in his introduction, and offers some possible reasons, including not wanting to make Plotinus' teachings public and the possible disrepute of Plotinus' reputation in light of the plagiarism charges (*VP* 18).

39. See, e.g., E. Lesky and J.H. Waszink (1959) 1229-32.

40. It is worth pointing out that both parents seem to be equal partners in supplying *logoi* to the offspring here. Elsewhere, however, Plotinus appears to describe the female as merely helping out in some passive manner (3.1.1.32-5; 2.3.14.31-2). Likewise, there is never any explicit mention of female seed, and both references to the *katamênia* make it out to be a passive principle (5.8.2.6-7; 2.9.12.18-21). Yet 3.6.19.17-25 provides further evidence for Plotinus' commitment to the female's active contribution. I suspect that the references to the female as a mere aid in the process are meant to refer not to the female *per se* but to her body, and that Plotinus throughout his career maintains that she has an active part to play in procreation. The only question is the manner in which she supplies *logoi*. See Wilberding (2008).

41. See e.g. Sorabji (2004) 1.211-2 and 3.362-3; O'Daly (1973) 38; Atkinson (1983) 228.

42. The term in question is *gonê* which need not be taken in the meaning of seed. See *AG* 9.4 and notes *ad loc.*

43. 4.3.12.37 and 13.1ff. Cf. also O'Daly (1973) 38.

44. 4.9.3.23-4; 4.3.7.25-6; 4.3.27.2-3; 4.4.32.9-11; 4.4.37.13.

45. cf. Galen *de Foet. Form.* 102,27-106,2 Nickel [= 4.699,3-701,6 Kühn]; and cf. *de Prop. Plac.* 90,18-94,17 Nickel (though see above note 12 on this text), as well as Themistius *ad Metaph.* 12.1070a26-8 (translated in Sorabji [2004] vol. 2, passage 1.15). I say 'would-be "spontaneously" generated' because once the World-Soul (or any soul for that matter) is said to govern the process of generation, it ceases to be genuinely spontaneous.

46. For what follows see also Wilberding (2008).

47. *AG* 2.4; 3.1; 10.1; 10.5; 14.1. Cf. *de Abst.* 4.20.35-6.

48. *AG* 16.1.2-3.

49. Porphyry *Sent.* §13. Cf. Plotinus *Enn.* 4.7.8^3.9-11; 3.8.5.24-5; 5.1.6.37-9; 5.1.7.47-8; 5.5.13.37-8; 5.8.1.19-21; 6.7.17.4-6.

50. e.g. *AG* 16.1.1-3 (55,30-2 K) and *AG* 10.5.1 (47,16 K), where *autê* refers back to *phusis* (10.4.10 = 47,14 K).

51. *AG* 14.3.9-10 (54,12-13 K).

52. *AG* 10.4-6.

53. See Long and Sedley (1987) vol. 1, 290-4.

54. One often finds such interpretations of Plotinus, e.g. Blumenthal (1971) 29; Tornau (1998) 282-3; Emilsson (1988) 26.

55. cf. Proclus *in Parm.* 791.38ff.

56. cf. *AG* 10.1.3-2.5 = 46,14-24 K; *Enn.* 5.7.2.

57. *AG* 5.4.4-9 = 41,21-6 K, and 6.1.13-15 = 42,15-17 K.

58. *Genesis* 30:37-9; Augustine *de Trin.* 3.5 and 11.2.5; Heliodorus *Aethiopica* 4.8; [Plutarch] *de Plac. Phil.* 5.12 (= *Moralia* 907E); Empedocles DK 31A81 (p. 300, *ll.* 19-22); Soranus *Gyn.* 1.39; [Galen] *de Hist. Phil.* 116 and *de Theriaca ad Pisonem* (in vol. 14) 253-4; Dionysius Halicarnassensis *de Imitatione* 17,18; Pliny *Nat. Hist.* 7.12; Aëtius *Plac.* 5.12.3 (= *SVF* 2.753). See also Bien (1997) 82-4.

59. cf. Heliodorus *Aethiopica* 4.119.

60. *de Caus. Plant.* 1.6.1-2 and 10. So too Clement of Alexandria *Stromateis* 6.15.117-8.

61. For an example in an ancient source of such influence in grafting, see *Geoponica* 10.76 (in Lenz (1856) 134) where, among other examples, mulberry scions grafted onto white poplar are reported to produce white mulberries. Compare this with Theophrastus' discussion of olives in *de Caus. Plant.* 1.6.10. For a modern account of graft chimeras, see Mohr and Schopfer (1995) 455-8 (*NB*: the more recent editions of this textbook are not as informative on this issue as the fourth edition).

62. *AG* 6.1. This is not to say that the mother has entirely replaced the father as the main causal factor. In *AG* 8.2 Porphyry seems to acknowledge that the father's contribution is slightly more significant. Cf. Simplicius' statement that the mother gives form (*eidopoiein*) to the offspring 'in a more proximate manner (*prosekhesteron*)' than the father (*in Phys.* 313,26-7). Cf. also Asclepius *in Metaph.* 404,17-18.

63. *GA* 730b19-32; 735a27-30; 738b10-21.

64. *AG* 10.3; 16.2

65. *AG* 13.6.8-10 (cf. 16.3).

66. *AG* 10.4.

67. But see *AG* 12.6.

68. 2.3.12.1-11, 14 *passim*, 16 *passim*; 4.4.38.19-23; 5.9.10.1ff. But cf. 2.3.17 where it seems to be the *logoi* themselves that are causing the problems, and 3.8.7.18 ff. where the culprit is likewise poor contemplation.

69. cf. *de Abst.* 3.27.38-40.

70. On the material make-up of the seed, see Plotinus *Enn.* 5.1.5.11-13; 5.9.6.15-20 (and cf. 3.1.1.35) and Porphyry *AG* 7.2; *Quest. Homer.* 8.583.

71. Theiler (in Harder, Theiler and Beutler (1956-71) *ad loc.*) sees *Enn.* 5.7.2.12 as a reference to the position in the womb as a determining factor in the offspring's features (cf. 4.7.5.44-5). On nourishment see *AG* 5.2; 10.3.

72. cf. *Enn.* 5.7.2.12: *hoi de en allêi khôrai pôs diaphoroi*; On Brisson's translation this passage is asking how different lands (*khôra*) help produce differences in children (cf. Plotinus' implicit suggestion in *VP* 11 that climate can affect one's bodily constitution even in later life). For another interpretation of this passage, see note immediately above. Note Igal suggests reading *hôrai* for *khôrai* here (see Henry and Schwyzer (1964-82) vol. 3, 325), which Armstrong (1966-88) adopts, so that the seasons rather than the locations that are at issue.

73. cf. *Enn.* 2.3.9.10-14 and 11 *passim*; 3.1.6.1ff.

74. *AG* 14.3.

75. *AG* 13.6.8-10 [53,14-16 K]; 16.3.4-5 [56,18-19 K]; 16.4.4-6 [56,24-26 K].

76. *AG* 16.3-5.

77. *AG* 16.5.1-3 [56,28-57,1 K]

78. cf. Festugière's cautious suggestion on 14.3 (cited in the notes of this translation *ad loc.*) that even here Plotinus means to suggest that the environment is active *after* birth.

79. *AG* 10.6.7-8 [48,3-4 K] and 11.4.9-10 [50,1-2 K].

80. For an excellent recent discussion of the text of the *AG*, see Dorandi (2008). Christian Tornau has kindly informed me that a PDF of this text, including introduction, indices, and images of folios 21^v and 23^v, has now been made available online by the Berlin-Brandenburgische Akademie der Wissenschaften at http://bibliothek.bbaw.de/bibliothek-digital/digitalequellen/schriften/ anzeige/index_html?band=07-abh/1895&seite:int=467.

81. Kalbfleisch dates the MS to the thirteenth century, but see Dorandi (2008, 123).

82. Kalbfleisch established that the 21 surviving folios, numbered from 3 to 23 (the first two pages are not counted as they are 'modern'), were arranged in quaternions, with folios 3-10 constituting one quaternion, folios 11-18 another, and 19-23 belonging to (but not exhausting) a third. Moreover, on the front of folio 11 there is a Greek number (probably *iê* [17] but possibly *ig* [13]) indicating that the quaternion which begins with folio 11 is either the seventeenth or the thirteenth quaternion of the manuscript. This means that there must have been 15 or possibly 11 quaternions (120 or 88 folios) preceding our first folio (f. 3). See Kalbfleisch Introduction page 3 along with his *Nachtrag zu S.* 3 on page 31, and more recently Dorandi (2008, 124).

83. See for example the *Vorwort* to Mau (1960).

84. On Minoides Mynas, see Omont (1916) and Gourevitch (2006).

85. 'Insofar as his [Mynas'] emendations and conjectures seemed right or at least noteworthy to me, they have been included in the edition with the symbol M; it is only with short and obvious emendations that I have not mentioned him for the sake of brevity, while he very seldom hit the mark with the longer ones' (Kalbfleisch 5n.1).

86. Three sections numbers (5.2, 13.5, 14.3) were inadvertently omitted from the text. Kalbfleisch acknowledges the last of these omissions in his Nachträge (80), and consequently this section division has been included in the TLG. By contrast the former two have not been corrected, and have been skipped over in the TLG's section numbering. (It is clear that 13.5 should begin at the full stop at 53,2 K. The missing 5.2, however, seems rather to be a case of faulty numbering, as 5.3 begins at the first full stop in the chapter). In conformity with the *TLG* I have included 14.3 in the section numbering of this translation, but not 5.2 or 13.5. The justification for this conformity is that their inclusion might confuse those using this translation in conjunction with *TLG* searches.

87. This is the first translation of *Ad Gaurum* to appear in English, though it is preceded by several translations into other European languages, two in German (Limburg (1975) and Jurisch (1991)). Neither translation, however, seems to me to be of much value compared with the excellent French translation by A.J. Festugière (contained as an appendix in Festugière (1953), vol. 3, 265-302).

88. cf. Sadler (2004) 47-8, 112 and 146-7.

89. *de Sem.* 92,19-94,11 De Lacy = 4.542-3 Kühn. Note that in his commentary De Lacy seems to assimilate these stages to the ones found in *de Foet. Form.* (e.g. *de Sem.*, *ad* 92,20-2), but see Nickel (2001) 126, especially n. 1.

90. In fact, the stages and vocabulary do not seem to correspond exactly to what one finds in *Nat. Puer.*, where one seems to find rather three stages: (i) seed [*gonê* – cf. 7.486,1 ff. Littré], (ii) flesh [*sarx* – cf. 7.92,18 ff. Littré], (iii) and child [*paidion* – cf. 7.498,27 ff. Littré]. Cf. Lesky and Waszink (1959) 1237 and De Lacy's note on *de Sem.* 92,28-94,1.

91. Galen *de Foet. Form.* 70,13f. Nickel (= 4.667 Kühn) with Nickel's commentary *ad loc.*

92. *AG* 5.3.5 [41,17 K]; 16.5.15 [57,13 K].

93. Porphyry only uses the participles. Similarly, in Galen the finite verb is also comparatively rare, not appearing at all, for example, in *de Foet. Form.*

94. Nickel's German translation of 'Keimling' works quite well but is unavailable in English.

95. 4.3.4. And cf. *zên phutikôs* 1.1.5 which is opposed to being a *zôion*.

96. Of course, there are other possibilities depending on the context: *dioikêsis* (9.1.5; 16.5.14; cf. 10.3.2); *zôê* (3.2.1-2); *energeia* (1.3.3-4); and cf. 4.8.12.

97. This is not surprising. Aristotle was equally flexible, cf. e.g. *DA* 415a23-5.

98. 10.6.8-9; cf. 16.4.1 and 18.1.2 where Kalbfleisch restores the text with *psukhê*. See also Porphyry's discussion plant souls in *de Abst.* 1.6.3; 1.18.1; 3.19.2.

99. The plural apparently indicates a belief in the theory of the *uterus bicornis*, which is to say the theory that the uterus has two 'horns'. See Lonie (1981) 114 and Nickel (1971) 82-3.

100. See Lonie (1981) 114 and 351. Plato also uses it in this sense once (*Laws* 792E3), but in the *Timaeus* he uses *mêtra* and *hustera* (91C1), as Porphyry realizes (*AG* 8.3).

101. The exceptions I have been able to find in Galen are all cases where he is citing Hippocrates, e.g. *de Foet. Form.* 56,3 Nickel (= 5.654 Kühn); *de Sem.* 76,14 De Lacy (= 4.525 Kühn).

102. Porphyry himself does not employ the other technical word for uterus, namely *hê hustera*, but he does cite Plato using it in *AG* 8.3. Here I translate Plato's *mêtrai te kai husterai* with 'wombs and uteri'.

103. cf. *AG* 2.5, 13.1.1-4 and 17.1. I do not think that Porphyry's comment at 10.1.1-3 is meant to mark a genuine division in the text to the effect that previous arguments were based on Platonic textual material and subsequent arguments are independent, as Festugière suggests with his headings ('Preuve par la doctrine de Platon, III,1-IX,5' [p. 270] and 'Preuve par la nature des choses X,1-XII,7' [p. 282]). Much of *AG* 3.1-9.5 does *not* depend on Platonic texts at all, e.g., the arguments in *AG* 3 and *AG* 7. Indeed, at 3.1.1 Porphyry emphasizes the *empirical* nature of the arguments to follow. Moreover, some of *AG* 10.1-12.7 does employ Plato (cf. 12.2). The comment seems rather aimed at signalling a contrast between what *immediately* preceded (the arguments in *AG* 9 which are very much grounded in the Platonic corpus) and what *immediately* follows (*AG* 10 where Porphyry presents his own embryological views).

104. cf. Plotinus 4.4.22.1-5; 4.4.27.1-7.

Porphyry
To Gaurus On How Embryos are Ensouled

1. (1) The judgment regarding the entrance[1] of souls into bodies in 33,1K/1.1.1
order to generate an animal[2] has filled not only us with great perplexity, Gaurus,[3] but even those who are primarily employed at examining this issue. Natural philosophers and nearly all physicians
are generally puzzled as to whether one should regard embryos as 33,5K/1.1.5
animals or say that they live in a mere vegetative manner, with the distinguishing character of animals consisting in sensation and impulse, whereas that of plants is observed to lie in a nutritive and growing power divorced from sensation and impulse. Hence, if embryos make do without representation and impulse and provide for themselves only in terms of nutrition and growth – for the facts bear witness to both of these things – one should agree that they are either
plants or similar to plants. And to hold that embryos are animals on 33,10K/1.1.10
account of their living in the manner of animals once they go forth from the belly[4] might well be hasty and something men do who are accustomed to go along with the opinions of the many without due
examination. (2) And more specifically, those who suppose embryos 1.2.1
to partake of an animal soul as well are again divided as to whether one should conclude embryos to be actual animals or merely potential
animals that are not yet actual. Potential in one sense is what has 33,15K
not yet received an ability but is able to attain it, as a boy can attain
the ability to read and write; but what has already received the 1.2.5
ability whenever it is not active with respect to that ability is potential in a second sense, as is the case with the boy who can read and write whenever he is engaged in other activities or sleeping[5] and is neither writing nor reading. Those who say that embryos are potentially animals predicate 'potential' of them not in the sense of [merely] being suited for ensoulment, rather they take it in the sense
of having already received soul without being active [with respect to 33,20K
its functions], since even those who hold that embryos do not yet
partake of animal soul would agree that they are potentially animals 1.2.10
in the sense of being suitable for ensoulment. (3) Is it the case, then, 1.3.1
that just as in a coma[6] the activities of the sensitive and impulsive [soul] are suspended even though the soul is present (as the release from the condition makes clear) or as in the case of hibernating
animals the vegetative activity remains slightly active throughout 33,25K
the period of hibernation whereas the sensitive and impulsive [soul]

34,1K/1.3.5 is completely inert, so too in the case of the [embryos] in the belly,
although the soul is present to the same extent that[7] it is once they
have gone forth [from the womb], their condition is like a coma or
hibernation? Or is the soul active but its activity is weak and can be
likened to the activity of using legs to walk, which babies are not yet
able to accomplish and yet they do move in accordance with repre-
34,5K sentation and bend and move around, even though they do not yet do
1.4.1 so by way of walking? (4) In the former case the [embryos] in the belly
are potentially animals, but in the sense in which things that have
acquired powers that are inactive are called 'potential' [with respect
to these powers]. In the latter case they are even actually animals.
And again, according to those who hold that they provide for them-
selves only in a vegetative manner and have no part of the impulsive
1.4.5 and sensitive soul, if they are said to be potentially animals it would
be said in the sense of being suited to attain the soul that makes one
34,10K an animal and surely not in the sense of having already received it
without using it.

2.1.1 **2.** (1) If, then, on the one hand, the embryo is shown to be neither
actually an animal because animals differ from non-animals by
sensation and impulse nor potentially an animal in the sense of
already having received a soul even though the activities of this
composite of body and soul are not being used, then the necessity of
[the soul's] entrance and determining when[8] this takes place become
34,15K/2.1.5 unproblematic for Plato. For clearly if the embryo is neither actually
an animal nor potentially one in the sense of having already received
the acquired power without using it; if it is rather said to be an
animal potentially by virtue of its being suited to attain the soul that
is properly said of animals when they first become capable of sensa-
tion and impulse, one cannot dispute either [the soul's] entrance or
34,20K the moment of its entrance, which must happen after it is naturally
2.2.1 born from the belly. (2) If, on the other hand, the embryo were
potentially an animal in the sense of having received the acquired
power, or worse yet if it were actually an animal, determining the
moment of the soul's entry will become problematic, and what is
more, whichever moment was determined to be the one will seem
2.2.5 implausible and arbitrary.[9] One man responds that it is the moment
when the seed is released since it would be impossible for the seed to
34,25K be fruitfully retained[10] in the womb unless a soul brought about this
natural merger (*sunphusis*) [of the masculine and feminine ele-
ments] by entering itself from outside.[11] In this group Numenius[12] is
a big name as are the exegetes[13] of Pythagoras' thought who under-
35,1K stand the river Ameles in Plato,[14] the river Styx in Hesiod[15] and in
2.2.10 the Orphics,[16] and the 'outflow' in Pherecydes[17] as indicating seed.
Another supposes that the soul enters when [the embryo] is first
formed which is in 30 days for a male whereas the female is articu-
35,5K lated in 42 days, as Hippocrates reports.[18] Still another determines

the moment of entrance to be whenever the embryo first moves.[19] Hippocrates has something to say about the time [of first motion] too:

> Whenever the extremities of the child's body sprout outwards and the hair and nails have taken root, this is when it also moves. And the time it takes to reach this point is three months for a male and four months for a female.[20]

(3) Once I heard someone earnestly maintain to us[21] that the male's eager desire during the process of impregnation and the sympathetic [response] of the womb[22] seize soul from the surrounding air through the breathing that is going on, transforming[23] the nature that was the seed's orchestrator[24] with the help of a characteristic power to draw in soul.[25] And [this soul] leaps out with the seed through the male as if through a pipe and is again seized by the eager desire in the womb whenever it is suitably disposed to retain [the seed and soul]. And this is why the two have intercourse because it is through both that the soul is bound and confined, and the experience was called 'conception' because these happenings are like trapping a bird.[26] But I laughed at these stories even then, and I know that I thought they merited remembering not because this fiction warrants any discussion but because this approach permits and reveals thousands of variations whenever someone refuses to place the soul's entrance after its birth from the mother and traces back this occurrence to the embryos that are still in the belly and to the obscure situation concerning that condition. (4) For those who think that Plato said this sort of thing, determining the moment of the soul's entry is problematic, and there will be at least an equal struggle for them when they try to show that the soul comes in from the outside rather than having a part of soul from the father be sown along with the seed (just as [a part] of nature [from the father is sown along with the seed]),[27] if indeed the soul comes in simultaneously[28] at the time of the sowing. And why isn't a part of the mother's soul sown along with it, if the soul arrives either when [the embryo] is formed or when it is first moved? For just as the similarities of its body reveal that it has inherited from its parents something from the body, so too must the similarities in soul indicate the source from which it has been received.[29] (5) And so our main concern shall be to demonstrate that the fetus[30] is neither actually an animal nor potentially one in the sense of having already received soul, and from this it follows that the soul enters after birth. And even if we were to concede that the embryo itself is potentially or even actually an animal, we will still maintain that its ensoulment can be neither from the father nor from the mother, but only from the outside, so that this account of the soul's entrance would not contradict Plato in this way, either.

3. (1) First, we resolve to establish the characteristic differences

between plants and animals by bearing witness to the very evidence
before our eyes, and then we resolve to examine in this way which of
the two comes closer to what happens with embryos. For if this
obviously resembles what happens in the case of animals, then we
36,15K/3.1.5 resolve to proclaim that the fetus is an animal, but if it resembles
what happens in the case of plants, let us not be surprised if it begins
its animal life once it has gone forth from the belly, just as we are not
surprised at how the seed, before it is separated from the father and
while remaining in itself, is inactive regarding those things that it
3.2.1 naturally creates once it hits the womb after secretion. (2) Well then,
it is peculiar to vegetative life and plants to nourish themselves not
36,20K through the mouth but rather through the power present in the roots
which draws the nourishing moisture lying about in the earth and
assimilates[31] it for the proper growth and nourishment of the thing
3.2.5 provided for by it. And again, it is peculiar to mortal animals of flesh
to receive their nourishment through the mouth and digest it within
themselves by means of organs built for this very purpose by their
36,25K/3.3.1 own nature. (3) And however many animals are of a nature to breathe
through their noses,[32] accomplish the inhalation and exhalation of air
through them. For when they are obstructed, they soon choke and
perish, not being able to endure, even for a moment, what is obstruct-
ing the continuous flow [of air]. Plants, however, transpire[33] only
3.3.5 through what is called the 'pith'[34] and their fruits are provided
36,30K nourishment and *pneuma* by means of what are called stalks from
which they hang and from which they fall after they have ripened
and received the final touch towards the culmination of their crea-
tion.[35] And in an environment that is on all sides saturated and
37,1K moist, land animals perish, whereas seeds issue forth from the earth
by their submersion in and storing and drawing upon the moisture.
3.4.1 (4) If, then, embryos also were nourished through the mouth[36] and
not through a power present in the seed – a power that draws in the
37,5K blood supplied to and submerging the seed within the womb (just as
plants draw moisture from the earth) and assimilates some of this
3.4.5 blood for the growth and nourishment of the embryo and discards the
rest of the blood as superfluous (though even this will be useful to the
finished product by reducing friction);[37] or if embryos did breathe
through their noses in the way they do after their advancement from
the belly and not through their umbilical cords (the membrane which
their midriff is suspended from and depends on in the manner of a
37,10K root or stalk, and from which they hasten to fall to the earth when
3.4.10 they are ripe in something like the way fruits do), or if they were able
to endure at least for a short time without being completely sub-
merged in moisture (whereas contrariwise the land animals after
birth cannot bare complete submersion for even a short while); if,
then, as I was saying, the administration of the embryos in the belly
37,15K did resemble that[38] of animals and did not at all resemble that of

plants, it would be possible for one who draws his beliefs from the 3.4.15
facts to agree with those who hold that embryos are animals. (5) But 3.5.1
since [the embryo] rejects the latter [*viz.* the animal's administration] as incompatible, while the one by which it is administered in a most appropriate manner is nearly contrary to the one it will have after advancing from the belly when it gives entrance to the animal
soul, why should we readily deceive ourselves by ignoring the evi- 37,20K
dence of the facts and concluding that what is inside the belly is an
animal simply because what we encounter at the moment of emer- 3.5.5
gence are animals? For either one must set aside those [facts] by which [embryos] are revealed to administer themselves as plants but not as animals, or else, since one cannot dispute the obvious, one must search for the causes of its transformation from plant to animal,
since this appears to be unbelievable to some and not a work of divine 37,25K
nature. However, one must let stand that the [embryos] in the belly are not animals, since they are equally immobile until they are ensouled as animals. 3.5.10

4. (1) But Plato, they say, says that the nature in the seed is a 4.1.1
part[39] of the appetitive, i.e, of the third part of the soul,[40] and that the appetitive part is guided by pleasure and pain and desires food and
nourishment,[41] and that pleasures and pains are sensations or at 37,30K
least motions that result in sensation[42] and that desires are im-
pulses.[43] Therefore, [they say], since embryos partake in sensation 38,1K/4.1.5
and impulse, by which animals differ from non-animals, how could
they not be animals? (2) But the men saying these things do not 4.2.1
realize that they are making plants into animals too, since even they are administered by a nature, which is a part of the appetitive part
of the soul. [Now] for us it is sufficient if these men should agree 38,5K
embryos live in a manner resembling that of plants but not that of 4.2.5
what are properly called animals, even if again in their eagerness they do abolish the difference between plants and animals by disputing the evidence and trying to deal with the problem at hand by means of an even greater absurdity,[44] since they have neither understood in what sense Plato says that the vegetative part is a part of
the appetitive part of soul nor are they eager to know the reason why 38,10K
Plato does not shrink from saying that plants are animals. (3) 4.3.1
Whereas others distinguish animals from non-animals through sensation and impulse, it is not the case that Plato likewise thinks that animals differ from non-animals by these [powers].[45] Rather, he makes the distinction between living (*zôês*) and non-living (*azôias*) reasonably and groups plants together with animals (*zôiois*), since plants are already living things (*zônta*). However, it is only nominally
that he grants [to plants] the common [status] with respect to what 35,15K/4.3.5
are properly called animals that partake of a self-moving soul. (4) 4.4.1
Presenting his own words will make both his own view and the error of those men clear. For he says:

Since all the parts and limbs of the mortal animal had formed a natural union, and since it so happened that this mortal animal[46] necessarily has its life in fire and *pneuma* and hence,
38,20K/4.4.5 being consumed by these and depleted, was decaying, the gods engineered some support for it. They begat a nature akin to human nature, but mixed it with other forms and perceptions so that it would be a different animal (*zôion*).[47] What are now domesticated trees, plants and seeds, were cultivated by the art of farming and made tame for us, though before there were only the species of wild vegetation that are older than the domesti-
38,25K/4.4.10 cated ones. For[48] everything that partakes of life (*tou zên*) might justifiably and most correctly be called an animal (*zôion*). Indeed, this thing that we are now discussing partakes in a third kind of soul which our account locates between the midriff and the navel; it has no part in opinion, calculation or intellection but does have a part in pleasant and painful sensation along
39,1K with appetites. For it remains always wholly passive, and its formation does not by nature allow it to perceive and reflect on any of its own features, by revolving in itself around itself, and
4.4.15 repelling external motion and exercising its proper motion. Therefore, it is alive (*zêi*) and is nothing other than an animal (*zôiou*), but it is fixed, stationary and firmly rooted, since it
39,5K lacks self-motion.[49]

4.5.1 (5) These are Plato's words, but those men who hold that according to Plato the embryo is an animal (*zôion*) – in the sense of 'animal' that pertains to our current investigation – run the risk of completely misunderstanding the philosopher. For if, on the one hand, someone wishes to call it an 'animal' (*zôion*) by reason of its being alive (*zên*), we shall agree; but we are positive that there is no way that Plato himself agrees that it has the self-moving soul that it has *when it is*
39,10K/4.5.5 *born*,[50] rather he thinks it is an 'animal' in the sense that plants are.
4.6.1 (6) What, then, is the sensation and desire of plants? Plato says it is homonymous and not identical to the sensation and desire of what are properly called animals. He points out this homonymy [when he says]: 'By mixing a nature akin to human nature with other forms
4.6.2 and perceptions so that it was a different animal, they begat'[51] the
39,15K domesticated trees and seeds. Plants employ other sensations and other desires and are 'animals' in a different sense than human beings. Therefore, even if according to Plato embryos have sensation, even if they have desire, even if they are called 'animals', it is only homonymously that they have these and are called 'animals'. However, [these terms are used] synonymously in plants, as we shall
4.7.1 show even Plato himself saying clearly. (7) Just as he in fact says that plants partake 'in a third kind of soul which our account locates

between the midriff and the navel; it has no part in opinion, calcula- 39,20K
tion or intellection but does have a part in pleasant and painful
sensation along with appetites'[52] and nevertheless, although they
partake in these things, they are not[53] animals in the sense in which
(*hôs*) the things that partake in self-moving soul are said to be 4.7.5
animals; rather, they partake only in a passive nature which they did
not shrink from calling 'soul', nor do they shrink from labelling the
movement conferred by this nature upon subjects 'life' in a sense
homonymous with what is derived from a self-moving soul. So, too, 39,25K
although embryos do partake of this same power and are 'animals'
according to Plato and called 'ensouled' and partake in sensation and
appetite, they are not animals in the manner in which the things that
possess self-moving soul are, and the investigation that we have 4.7.10
presently undertaken is aimed at discerning when this soul enters
into the body. (8) One must not get confused by the homonymy and 4.8.1
make mistakes; rather, just as although Plato says that plants are 39,30K
cultivated by the art of farming and made tame for us, though before 40,1K
there were only wild ones,[54] we do not because of this suppose that
intellection and calculation and character belong to them as one
really speaks of 'cultivation', 'taming' and 'wildness' in connection 4.8.5
with such things. Instead, we understand the latter terms meta- 40,5K
phorically or even homonymously. So too must one understand sen-
sations, desires and appetites to be meant homonymously and in the
manner of an analogy and not to be the ones provided by the self-mov-
ing soul. It is this soul's entrance into the body that our account has
been about, and we agree that that which is called 'soul' in a different 4.8.10
sense, as being subject to strokes[55] and affections,[56] and in which
plants partake, is also what embryos partake in, but that even
according to Plato they have no part in the self-moving soul prior to 40,10K
exiting the belly, even if they are participants in highest degree of
vegetative sensation and pleasure. (9) For although others[57] rank 4.9.1
sensations under the impression that comes to be in the soul by
means of the sense organs when the soul is attached to the body,[58]
our man [*viz.* Plato] seems not to do that at all. Rather, he seems to
call the corporeal motion in which even plants partake 'sensation'
and to label the motion of the soul that is yoked together with this 40,15K/4.9.5
corporeal motion 'opinion'. And he seems to constitute the apprehen-
sion of perceptibles out of both the irrational, passive motion – which
he labels 'sensation,' not 'motion of sense organs' – and from opinion,
which others suppose to be an impression in the soul. And this is why
when Plato defines 'perceptible' he says '[it is] what is grasped by
opinion with irrational sensation'.[59] (10) Accordingly, since Plato does 40,20K/4.10.1
not lump together the strokes that do not involve understanding,
knowledge or representation with the judgments that belong to this
opinionative soul, that is the cognitive soul, and places the former
even in plants,[60] he reasonably says that plants partake in sensation.

And he might be offering a summary of the doctrine on these [*viz.*
plants] not partaking of the soul that is self-moving and at once both
4.10.5 opinionative and reasoning and also – as others would say – respon-
40,25K sible for sensation and impulse, when he says: 'For' the plant 're-
mains always wholly passive, and its formation does not by nature
allow[61] it to perceive and reflect on any of its own features, by
revolving in itself around itself, and repelling[62] external motion and
exercising its proper motion'.[63]

4.11.1 (11) And one might say the same thing about embryos, and still
more so what [Plato] adds about plants, since this similarly goes for
41,1K embryos, too. He says: 'Therefore, it is alive and is nothing other than
an animal, but it is fixed, stationary and firmly rooted, since it lacks
self-motion'.[64] For the embryo, too, is rooted by the umbilical cord and
4.11.5 remains inside and is called an 'animal' (*zôion*) by reason of its being
alive (*zên*), but it does not yet partake of the self-moving soul.

41,5K/5.1.1 **5.** (1) But they say that embryos both move locally and receive
sensation of burning heat because they jump whenever the mother's
belly encounters the burning air in the baths, and others have still
more generously posited that embryos even have the bizarre appe-
tites to which the mothers are subject throughout the duration of the
pregnancy since the mothers neither had come to experience them
41,10K/5.1.5 [*viz.* the appetites] previously nor did they encounter an affection
similar to these after giving birth. And this [they say] is shown by the
fact that [the mothers] bring forth offspring unharmed by satisfying
these appetites, whereas by not successfully obtaining the objects of
these appetites they bring forth offspring bearing indications and an
5.3.1 impression in the body of how the desired things were lacking.[65] (3)[66]
And last but not least, they claim that the process of giving birth
itself reveals that [the embryo] shares in the [power] responsible for
impulse. For [they claim] still-born deliveries are painful since [the
41,15K foetus] is not contributing any impulse to nature for its delivery, and
the delivery of female offspring is more time-consuming, since [fe-
male foetuses] are sluggish,[67] and the impulse of the mother is not
5.3.5 sufficient to push the embryo out unless there is also an impulse to
5.4.1 get out contributed[68] by the offspring. (4) But if just this once it were
necessary to yield to this nonsense, then *I* would help them find a way
to say that embryos also have representations and opinions together
41,20K with the mother, so that they would partake of the soul responsible
for representation and opinion. For there is a consensus that many
animals and in particular women in the act of fertilization, when they
internalize with their [faculties of] representation forms coming from
5.4.5 the same species,[69] bring forth offspring that are very similar to those
forms. Therefore, we place in front of horses, dogs, doves and indeed
even in front of a woman, images that realize a beauty of form, since
41,25K copulating females, by looking at the appearances and receiving
them into their memory, bring forth offspring that resemble them.

(5) Now it is easy to claim that that could not have happened unless the seeds partook of the soul responsible for representation. For how, when one thing is engaging in representation, could another thing with no share in representation be moved in accordance with the state of the representing thing? For this would be just like if, while you were undergoing some affection and I wasn't, I were still to be affected on account of my being enclosed in the same room or, if you prefer, by my being chained to you.

6. (1) But all of these [arguments] are due to the power of abstract argument (*logos*)[70] which can easily supply an abundance of false and misleading claims as well as eliminate the truth by appealing to plausible likelihood. For example, to begin with the last [points],[71] if we were able to make whatever we represent rub off onto our own bodies in the manner in which – and here already abstract argument [as opposed to fact-based reasoning] is holding sway – *daimons* display the forms of [their] representations in the airy *pneuma* that either accompanies or is adjacent to them; and without touching the *pneuma* in any way, they nevertheless display – in a way that cannot be described – the images (*emphaseis*) of their faculty of representation by means of the air around them as if in a mirror;[72] [if we could do this,] then it would be possible to infer that there is a faculty of representation belonging to the soul present in the seed which is what shapes the body after itself. However, since we are not able to do this to ourselves, though we are able to form other things according to our representations, namely whatever is external to our own substance, perhaps for this reason a soul proper to the embryo is not the craftsman of the formation of the [body] subordinate to it. Rather, it is the mother's soul that – though not being the craftsman of her own body either – is the craftsman of someone else's body which is in the mother and yet external to her substance, since even in other cases the mother's soul is able to make representations rub off on things external to her substance. (2) But one should realize that this too is certainly going to be said in conformity with Plato. For according to him[73] the things that have been engendered from the substance of some things are always a step down from the things that had engendered them in terms of power and substance;[74] and it is impossible for them to be of the same substance as the things that engendered them, yet they do in a way become obedient to those that brought them forth and are perfected by them. For in this way discursive thought which is an offspring of intellect is substantially a step down from this intellect that engendered it, and yet it is able to turn back toward intellect and understand its content, even though, unlike intellect, it is not party to a non-discursive contact that takes place all at once.[75] Again, the irrational power contiguous to reason, being itself an offspring of reason, is in substance not party to rational thoughts, but it is said[76] to be in accordance with reason,

and although in terms of its own substance it is unable to move
6.3.1 rationally, it is perfected by reason. (3) And the vegetative [power],
being an offspring of the irrational power and in general of the
appetitive [part of the soul], is according to Plato substantially a step
42,30K down from the soul responsible for opinion and representation, but it
is able to be maintained by the soul responsible for representation
43,1K even though it is not party to representation and opinion. In this way,
6.3.5 at any rate, even plants are said to be cultivated and tamed by
farming – not because they are receptive to the farmers' voice by
means of representation – but because they are able to be led and
43,5K steered by a conductor, being directed by means of their affections.[77]
6.4.1 (4) Thus, it is not at all surprising that the vegetative [power] of the
female, once it has formed a natural unity (*sumphusan*) with the
power of the seed, while maintaining its obedience to the part of the
mother's soul responsible for representation, is affected by the form
of the object that is jointly represented.[78] For we also said[79] that
sensible affection is proper to what is disposed to be affected, whereas
6.4.5 understanding and knowing related to it are proper to the self-mov-
43,10K ing soul. But the form of the thing being shaped depends on affection
and impression and not on understanding and knowledge.
7.1.1 **7.** (1) And how could it be owing to impulse and representation
that spatial motions belong to embryos which are rooted by the
umbilical cord, when these motions are rather like the turning
motions of bowels and the vibrations of organs in cases of flatulence?
7.2.1 (2) For even in us there arise very many spatial transmissions of
43,15K things that do not involve representation. An example is the way that
the transmission of nourishment is explained. For the belly's recep-
tion of nourishment proceeds consciously at first as long as [the
nourishment is] between one's teeth and one's throat, but from then
43,20K on there is no consciousness – neither of how [the stomach] concocts
7.2.5 [the nourishment], sending the useful portion up to the liver and the
waste to the rest of the [gastric] cavity and the bowels, nor of its
sending the liquid waste to the bladder. Nor again can one possibly
have a representation of [the nourishment] changed into blood in the
liver or of how nature sends [the blood] up to the heart having
separated off the thick portion into bile, or how the heart sends [the
blood] back through the veins and these moisturize the skin. Nor is
it possible to perceive how [nature][80] extracts some part of this
[blood][81] and turns it into seed by reproducing its own form principles
7.3.1 in the thing coming to be. (3) But all of these, although they are
43,25K spatial motions, are not accomplished through impulse or repre-
sentation, just as the motions of embryos are not either. Further,
both pleasures and pains are expansions and contractions,[82] which
7.3.5 are observed even in plants both when they wither due to thirst and
when they sprout afresh due to the taking in of moisture. Hence, just
as these plants are said to be thirsty or again to be sated without

having representation, and just as some plants[83] turn toward and 43,30K
revolve with the sun, moving together with it by bending towards the
arc of the sun, while other plants[84] unfold towards the moon and open 44,1K
up and greatly extend themselves [in bloom], still other plants[85]
stretch their tendrils out like hands towards their vine-props, and in 7.3.10
this same way some [plants] also sway in a natural manner[86] towards
heat.

8. (1) And saying that the mother desires what the embryos for 8.1.1
their part are presumably desiring is characteristic of men who are
completely ignorant of the events that happen to pregnant women on 44,5K
account of their womb. Rather, just as one must not say that the
pregnant women vomit because the [embryos] in the belly are vomit-
ing, nor again that they are nauseous on account of the nausea of the
embryos, so too must one avoid saying that pica is due to the embryos' 8.1.5
appetites; rather, one must attribute the entire affection to the
motion of the womb which also brings about the affection of blows in
the fetus when [the womb's] appetites are unfilled.[87] (2)[88] [Plato], of 44,10K/8.2.1
course, thinks that the womb is steered by its own impulse and
grants to the womb a role in the creative process nearly equivalent
to that of the part given by the father to the created product. And
these are his exact words in the *Timaeus*:

> the gods contrived the love of copulation, constructing the one 8.2.5
> ensouled animal in us [men] and another in women[89] 44,15K

Plato calls the male member and the female womb 'animals' in us
through which generation proceeds, but they are 'animals' (*zôia*) not
in the sense of living (*zêi*) merely as plants do, rather in the sense of
being obedient to the self-moving soul too. For he says:

> the nature of the genitalia is unruly and self-willed like an
> animal that does not obey reason, and tries to dominate every- 8.2.10
> thing owing to its raging appetites.[90]

The [term] 'self-willed' (*autokrates*) makes clear that motion arising 44,20K
in the genitalia is obedient to impulse, and what happens makes this
clear as well. For they are led by representation, the very thing by
which the other parts that are obedient to impulse are led by, too. (3) 8.3.1
And Plato writes the following about the womb:

> And again the wombs and uteri in women are for these same
> reasons called an animal within them with an appetite for
> producing children. And whenever it is barren for an unseason- 44,25K
> ably[91] long time, it becomes severely irritated and wanders
> throughout the body,[92] and it blocks up the passages for the 8.3.5
> breath (*pneuma*) and prevents [the woman] from breathing,

sending her into great difficulties and presents her with all
kinds of illnesses, until the [female] appetite and the [male] love
draw them to each other, like stripping fruit from the tree, and
44,30K sow into the uterus as if into fertile soil animals not fully formed
and too small to be seen, and again once they have given them
8.3.10 distinct form, they nourish them large and after this, by bring-
ing them to light, they complete the generation of the animals.[93]

8.4.1 (4) And so according to Plato even the womb is 'an animal with
45,1K appetite' and it is 'severely irritated' and 'wanders throughout the
entire body'.[94] How, then, is this not the thing responsible for the
appetites [of the mother] and motions [of the embryo], seeing that
Plato even explicitly says that it 'send[s] her into great difficulties'?
Therefore, both pica and the motions of the embryos have their
source in the womb.

45,5K/9.1.1 **9.** (1) For the intellectually capable to gain knowledge of Plato's
opinion, it is sufficient for him to have said that the appetitive powers
in the reproductive parts of the parents pluck the seed – like fruit
from trees – and sow it in the womb as if in fertile soil.[95] This is
9.1.5 sufficient for them to have a clear understanding that <the>[96] ad-
45,10K ministration of the embryos in the womb is vegetative according to
9.2.1 Plato and does not yet involve the self-moving soul. (2) They, how-
ever, ask how Plato can say that they sow '*animals*'? Yet these men
are deceiving themselves by disregarding a great deal. For Plato does
not simply say that they sow animals, although even if he had said
this it would be possible to understand him in the same manner we
did a little before when he said that plants were animals. But what
9.2.5 *does* he say? They 'sow animals unformed and too small to be seen'.[97]
45,15K But what is unformed is not yet an animal! And even if the embryos
had been formed and nourished,[98] nevertheless the figure belongs, we
agreed,[99] only to the body and the nourishing proceeds by a vegeta-
tive power and not by the other soul, which is strictly speaking that
of an animal. 'And after this', [Plato] says, 'by bringing them to light,
they complete the generation of the animals'.[100] But they complete
9.2.10 the generation of the things that *are able to become* animals. And in
this way Plato is found to have a clear understanding that the
45,20K ensoulment of the self-moving soul occurs after [the embryo] has
9.3.1 gone forth from the mother into the light of day. (3) But how, they
say, can Plato at one point, while going over the misfortunes of the
soul, say 'if they [the souls] are indeed laid up in store and will come
to belong to the things carried in the womb'?[101] But those who say this
do not realize that there is a difference between 'being carried in the
womb' and 'coming to belong to the things that are carried in womb'.
For the one means that something is itself being put into the womb,
while the other means that [something] is being introduced into the
45,25K/9.3.5 things that are [at some time] carried in the womb. And so just as

when someone says that coming to belong to mortals is a misfortune of the soul, he does not mean that [the soul] comes to be mortal but that it is introduced into mortals, so too must one presumably understand 'belonging to the things that are carried in the womb' to mean that [the soul] comes to be in *the class* of things that are carried in the womb, that is to say, in the class of mortals, and not that it [the soul] itself is being put in the womb together with them at the very moment when they are put into the womb. (4) And they similarly 9.4.1
misunderstand Plato when he says in the *Phaedrus* that the soul that has seen the most will go into the *gonê*[102] of some man who is musical 46,1K
and a lover of beauty.[103] For one must understand *gonê* to mean not 'seed' but 'generation' so that it refers to a generation of the man who is still to be put into the womb [*viz.* conceived] and not a seed of the man who is begetting. For Plato says that [this soul] comes and is 9.4.5
introduced not into the seed [emitted by] some lover of beauty but rather into the generation of a man who loves beauty, so that the lover of beauty is not the one who brings forth offspring but the one 46,5K
who is engendered. (5) And it is made clear in the *Timaeus*, also, that 9.5.1
according to Plato the soul is introduced after the completion of the body when [the embryos] go out from the mother into the light of day. For he brings the soul into the entire body after it has been built.[104] And likewise in the *Phaedrus*[105] and in nearly all those places where Plato says that it is the soul's taking hold of the body that brings 46,10K
about the composite as an animal.[106]

10. (1) It is also necessary to put Plato aside and investigate what 10.1.1
happens all by itself and to consider with no less dedication whether the entire generation of embryos is as much as possible like that of plants, with the father emitting the seed and the mother collecting it in order to nourish it – not merely in the manner of earth for the 46,15K
supply of nourishment nor in the manner in which she furnishes milk 10.1.5
alone to offspring after birth – but rather resembling in a way [buds] that are being grafted and added onto [the base of another plant]: <the>[107] power in the womb forms a natural unity with the seed, and it is by this power that the base [*viz.* the womb], which has its own nature, and the part being added on [*viz.* the seed], which has its proper nature, make up a kind of mixture in order to form the single nature of the part that has been successfully added on [*viz.* the embryo]. (2) Then, the administration of the engendered thing pro- 46,20K/10.2.1
ceeds in accordance with the grafted part, but whatever [functions] naturally come from the base (*katôthen*) are administered by the part [*viz.* the womb] that has received [the grafted part] and in accordance with its own nature. And sometimes the qualities of the base predominate, and sometimes those of the grafted part take hold of the whole.[108] (3) Indeed, both in plants and in the vegetative womb the 10.3.1
administration is nearly the same: the power in the seed immediately 46,25K
puts a membranaceous shell around its exterior, as Hippocrates

says[109] – just as in the case of fruiting shoots[110] it completes[111] the
flower and the pod – and this[112] becomes the chorion, and [this power]
10.3.5 draws out a thin tube, like an intestinal tube, from the middle, in the
manner of a root or a stalk, and the embryo, hanging from and rooted
by this, draws breath[113] and most of all supplies itself[114] with nour-
46,30K ishment from it.[115] This tube they call an umbilical cord[116], and from
the spherical circumvolution of the seed in length and breadth [this
47,1K power] stirs up another external membrane,[117] which becomes for the
thing being formed its defence against exterior [dangers],[118] and in
10.3.10 the remaining time up until [the embryo] goes forth [from the womb]
[this power] forms all the interior parts and makes them solid. At
least, if a premature [embryo] is drawn out of the womb by force, even
if it is nearing the time of parturition, one will discover that its
insides are dissoluble and not firm, although the shape is present and
the exterior shell has been manufactured.

47,5K/10.4.1 (4) Indeed, the entire time in the belly is spent on forming and
solidifying, like in the building of a ship, where at the very moment
when the ship-builder has completed the ship and launches it into
the sea, the ship's captain (*kubernêtês*) is introduced into it. And if
10.4.5 you imagine for me that the ship-builder has always been joined
together with the ship and does not withdraw from it when the
47,10K captain of the ship, which has been launched from land into the sea,
boards it, you will have an image resembling the formation in the
generation of an animal,[119] but in many other respects the works of
nature differ from the hand-crafted products of the ship-builder –
especially in that the ship-builder is able to be separated from both
10.4.10 the product of his creation and the ship's captain, whereas nature is
47,15K inseparable from the things that it makes and always wants to enter
10.5.1 completely[120] into its works. (5) Therefore, nature even joins with
different captains at different times: (i) For as long as the seed is in
the father, it[121] is administered by the vegetative [power] of the father
as well as by the father's soul from above which conspires with the
vegetative power towards its works. (ii) And once it has been released
from the father into the mother, it joins the vegetative [power] of the
47,20K/10.5.5 mother and her soul – and by 'joins' one must understand, not that
both are destroyed together nor that they are resolved into their
elements like <un>mixed items,[122] but rather that they maintain
that divine and paradoxical [kind of joining], blending, that is the
special power of living [substances]. And so in this way [they] are
both united with the suitable things just like elements that get
destroyed [*viz.* by uniting with other suitable elements] in mixtures,
and again preserve in this way their own powers just like things that
47,25K/10.5.10 are unmixed and are separated out by themselves.[123] And this is
indicative of their neither being bodies nor even having their sub-
10.6.1 stances because of the dispositions of bodies.[124] (6) However, as far as
concerns the complete blending which does not entail the destruction

of its parts, I am prepared to provide an account of the appropriate length in other sacred books.[125] (iii) And once [this power] is no <longer> under the direction of the mother,[126] and indeed when the mother does not resist the [force] that will cut off the blending *via* separation, that [power] for its part is delivered by the laws of nature from darkness to light, from its abode moist with blood to the airy hollow. And it in turn at this time immediately gets from outside the captain who is present by the providence of the principle that administers the whole, which in the case of animals would in no way let the vegetative soul come to be bereft of a captain. By contrast, the fathers, when this vegetative power is with them, only fortify it towards its task, whereas the mother's external soul, by infusing it, fortifies it toward its task, just as the fathers did, *and* steers the task, just as human souls <steer> the tasks of their own natures.[127]

11. (1) When the nature [of the offspring] has gone forth into the light, the captain steps in [to deal] with the task, without being compelled to do so.[128] And just as I have seen [happen] in theatres, the actors who play Prometheus are compelled to make the soul enter into the body once the formation [of human kind] has been established.[129] The ancients perhaps wanted to point out through this story that there is no compulsion in the [soul's] entrance [into body], by pointing out that it is only after birth, once the body has been formed, that ensoulment takes place. Indeed, even the theologian of the Hebrews seems to indicate this when he says that once the human body has been formed and received the bodily creation in its entirety, god breathed *pneuma* into its living soul.[130] (2) Thus, when the self-moving soul goes into the bodies, it is not being compelled to do so and still less does it have to aim for the mouth and nostrils.[131] Some Platonists are proud of these ridiculous tales that one should be ashamed even to mention.[132] For the ensoulment is natural, and kindling[133] in general[134] [proceeds] according to the harmony of the [things] joined together with respect to their capacity for being joined into one. For it is in this way that even an eye sees – not because a cone has been extended from the pupil to heaven,[135] nor because a diffusion of rays to the object of sight produces angles by reflection,[136] nor because images flow out of the things seen to the seer[137] – but rather the one sees and the other is seen because they have joined together in a suitable manner.[138] In the way that even naphtha[139] has been seen to be kindled by fire all at once[140] without the use of a medium[141] as long as nothing gets in the way so that the well-fitted [parts] will experience the same things as one another even without touching.[142] And the magnetic stone naturally draws iron and <amber>[143] [naturally draws] small bits of wood on account of their kinship, and what is well-fitted for the soul's steering [draws] the soul that is suitable to the well-fitted thing. And neither willing nor aiming nor choosing contribute anything to its presence, since if the

harmony is dissolved not one of these keep the soul from departing.[144]
11.3.1 (3) And so, just as [in death] when the instrumental body comes to be
49,5K ill-fitted – even if someone employed thousands upon thousands [of
devices] to block up the mouth and nostrils and all the body's other
cavities,[145] even if someone should *choose* to have the soul remain or
force or *beg* it to remain, it goes away by the necessity of nature which
11.3.5 ordained that two ill-fitted things in no way form a harmony; so too
[at birth], when the body has become suitable for receiving [the soul],
49,10K the soul that will use the body is present without having to be
implanted in it little by little or having to enter [the body] through
one of its parts. Rather, its presence and dissolution are instantane-
ous, without extending its existence by having passed through a
process of generation or destruction – in the very same manner that
11.3.10 lightning's existence is not caused little by little through a process of
generation but either is or is not without extending its existence
through the process of generation or destruction. Nor is it the case
that the soul soars through the air and flies through the mouth or
49,15K nostrils like a bird entering a house through a window; rather,
regardless of whether the soul, because it is part of this [sensible]
world,[146] drags behind itself some body of aether or *pneuma* or air or
11.3.15 some combination of these,[147] or whether it is able to establish its
presence in animals that are suitably disposed without these [vehi-
cles], the ensoulment of the underlying [body] takes place all at
49,20K once,[148] just as the rising of the sun is instantaneous[149] as far as its
distribution of rays goes which extends from one end of the earth to
11.4.1 the other and includes everything made visible by the sun. (4) And
just as strings[150] that have been tuned to the same scale, though they
be separated by a great distance, if wood-chips have been placed upon
these [distant] strings as well as on strings nearby – only let the
strings nearby be not in tune [with the others] – when one of the
tuned strings is struck, [the distant strings that are in tune] vibrate
49,25K and shake off the wood chips while the strings nearby are still and
remain unaffected due to their not being in tune; and the distance
11.4.5 between the tuned strings was in no way obstructive to their unity of
affection nor did proximity overcome the lack of sympathy of the
untuned strings. In the same way, the instrumental animal that is
well-fitted for a suitable soul immediately possesses in sympathy the
soul that will use it. And the soul's sympathy for this [body] but not
50,1K that [body] is granted by its previous existence or even by the rotation
of the universe that leads like to like.

12.1.1 **12.** (1) If, then, someone is able to show that the body is already
well-fitted during its state of incompletion in [the womb], one would
have to posit also that the soul that will use the body suited for its
use comes to be inside [the womb] as well. But if the body, when it is
50,5K still being carried in the belly, is not set – if there is still a need for a
maker and a tuner of strings, but not for a player of strings – why

would we refuse to face the facts and expose ourselves to serious 12.1.5
absurdities by thinking that just because we did not observe the soul
entering at birth, embryos emerge having already received this soul
beforehand? And, we say, they can be observed not to possess intel-
lect in their childhood either. For intellect, too, certainly enters as 50,10K
one's age advances.[151] (2) I, for my part, do not refuse to endorse this 12.2.1
view, as if I thought it were bizarre;[152] rather I would affirm this view
above all else. And I am bringing in Plato and with him Aristotle as
witnesses to the position that intellect comes to men late and does
not even come to all men in this way; rather, rare is he whose soul
becomes suitable for union with intellect.[153] (3) But if when soul turns 50,15K/12.3.1
to itself it discovers that this intellect is present, it is surely not the
case that, on account of this, the soul also originally was in possession
of intellect when it entered [the body]. For there are things that are
mastered but in no spatial manner and are held together but in no
bodily manner, rather they form a natural unity through the suitabil-
ity and likeness of the things on the receiving side.[154] And it is neither
place nor time nor any other force that holds these things together. 12.3.5
Rather, it is lack of suitability that obstructs and dissolves [their 50,20K
unity], and suitability that brings them together, masters them, and
holds them together for however long the harmony between them
exists. This also explains why he who knows god has god present in
his possession, while he who is ignorant [of god] is absent to him who
is present everywhere.[155] And there is no reason to fear that someone
might censure our taking intellect away from children,[156] regarding
which both Plato concluded that a man should be content to have it
arrive in old age[157] and that the souls come into generation after they 12.3.10
have shed their wings, [158] and Aristotle demonstrated that in whom- 50,25K
ever intellect arises, it arises from without.[159] And the soul arrives
having calculations and choices and opinions from the start, but the
human being who is an infant is still incomplete.[160] (4) And the life in 12.4.1
the belly is not like the incompleteness of an infant compared to an
adolescent; it is rather of an altogether different character and sort.
For as time goes by, [the infant] will receive dry food, whereas in its 51,1K
infancy it is nourished by milk, and milk does not lie outside of the
class of real foods,[161] nor does the nourishing [proceed] through 12.4.5
anything other than through the mouth. By contrast, the nourishing
that takes place [with the embryo] in the belly is peculiar and
[proceeds] not through the mouth but through something else –
namely the umbilical cord[162] – and this [manner of nourishing] be- 51,5K
longs rather to plants and not animals. And the infant utters sounds
without any sense or meaning, but it nevertheless signals what
causes it pain through gestures and crying, whereas in embryos the
way of life is silent and does not involve representation. And in other 12.4.10
ways as well (lest we work through the same things again) the
nourishing of the embryos that are still being carried in the womb

51,10K/12.5.1 comes closer to the manner of life of plants than that of animals. (5)
Now, the rational part of the soul exists in a way even in babies
because the part of soul responsible for sensation is present in them
and it is through this that the rational part is roused to actuality; but
nevertheless in order to achieve actuality it requires a significant
contribution from those who live with it, i.e. an external contribution
from the rational part that is already in a state of actuality.[163] By
12.5.5 contrast, in the [embryos] still being carried in the womb there is not
even a trace of the part of soul responsible for sensation because they
do not even have the organs through which sensations proceed[164] and
because there is not yet the space nor the distance [required to
51,15K/12.6.1 perceive] in there.[165] (6) As for palpitations, in superfetations[166] they
are not fewer, and the sort of turning motions that belong to embryos
are found in these cases as well,[167] so that both midwives and even
the mothers themselves in the throes of childbirth have often been
deceived because the thing in motion was something that has been
generated alive. And the condition that is called having a mole also
12.6.5 presents the same movements and palpitations so that it becomes
51,20K unclear for a long time whether what is in motion is not also an
embryo, but once it has outlasted the duration of pregnancies it is
exposed [for what it is].[168] But these things [superfetations and
moles] are agreed not to have souls, even though they are not without
part in a nature.[169] Monstrosities are of this sort, too, since they –
although they are unnatural – are [still] in the possession of a nature
that has failed to achieve the offspring that corresponds to its form
12.7.1 principle.[170] (7) At this time there would be need of the nutritive
activity of growing and formation, but the arrival of the power
51,25K responsible for sensation and opinion would be superfluous as well
as obstructive. For even now[171] it is the case that if nature is going to
do the nourishing and finish the work that concerns it without
12.7.5 obstruction, sensation is obstructed – it wanes during periods of
sleep[172] – and to whatever extent the sleep is free of dreams and
representation, nature is to this extent more active,[173] while those
who are sleepless and full of anxiety are poorly nourished, since the
work of the one [power] is frustrated by the activities of the others.
51,30K If, however, there is also need of a faint echo of sensation – as if for
light when no sun-beam comes in from outside – it is sufficient for the
52,1K creation to be completed in an underlying thing that has sensation,
12.7.10 namely in the body (*onkos*) of the mother.[174]

13.1.1 **13.** (1) These considerations are sufficient to show that the embryo
is not actually an animal and that it does not partake of a self-moving
soul that is actually present. But it remains to show that it is not even
52,5K potentially ensouled, where 'potentially' is understood to mean what
has received the state but is inactive and not acting. What has
13.1.5 attained the state and does not act but rather remains inactive in its
potentiality, is complete with respect to its form <and> merely still.[175]

If it is incomplete with respect to its form and one [still] predicates
'potentially' of it, one is obviously switching to the other sense of
'potentially' and rejecting the sense of 'potentially' which is applied 52,10K
to the complete form which is merely at rest without acting. (2) For 13.2.1
in the same manner ...[176] he who predicates 'potentially' of the oar
which has been thrown from the ship is discovered to mean 'poten-
tially' in the sense of being in accordance with a complete state while
refraining from its activity and remaining in its proper potentiality,
on account of the fact that the oar is not propelling the ship. But he
who applies 'potentially' to pieces of wood that are suitable for being 52,15K
assembled into an oar where the form of the oar is not yet visible in 13.2.5
them but can be generated by the skill of a carpenter, agrees that the
form of oar does not yet exist in this wood at all. With the designation
'potentially' he means rather that it is able to become [an oar], so that
the one [i.e. the pieces of wood] is said to be 'potentially' [an oar]
because of its capacity to receive the potentiality whereas the other
[i.e. the lost oar] is said to be 'potentially' [an oar] because it remains 52,20K
inactive in the potentiality that has already come to be in it. (3) 13.3.1
Further, since one who is sleeping is not employing any of the
activities related to sensation and the fetus is not employing them
either, one must examine whether either the seed or that which is
created from the seed is like an inactive man who is sleeping or even,
by God, one who is comatose[177] – and whether [it is like] the oar that
is inactive and not propelling the ship rather than being like the wood 52,25K
which has not yet attained the form of oar. (4) And surely the embryo 13.4.1
appears to remain incomplete until parturition to him who investi-
gates the truth [of the matter]. For the animal body is not the
external formation, and the form is not the shape that surrounds its
exterior. It is rather the full completion of the formation – the
perfection of its external shape as well as of its inner parts and
everything else, the sinews and bones, the arteries and veins, i.e. the 52,30K/13.4.5
observed completeness of the production of all its organs,[178] just as
with fruits grown in the upper branches of trees, the ripe fruit has
achieved its completion, with embryos the one that naturally departs
from the mother, like fruit falling from a tree, is ripe.[179] Therefore the 53,1K
embryo is not yet sensitive in the sense of having received the state,
and therefore neither is it potentially [perceptive], at least in this
sense. For the sensitive powers are not yet present in incomplete 13.4.10
organs, just as the form of oar does not yet exist in pieces of wood.
Therefore, the embryo is potentially sensitive and desirous and 53,5K
reasoning – not in the sense of having received [these powers] and
being inactive in its potentiality, because it is not like someone asleep
– but rather in the sense that it is *able* to receive [them] and is still
incomplete. (6) Plato, too, is of this opinion when he divides the soul 13.6.1
and assigns the control centre to the head, the spirited part to the
heart, and the appetitive part to the liver.[180] And how can there be a 53,10K

13.6.5 place for the soul to enter where there is not yet any head or heart or
liver?[181] And if nature – and still less God – makes nothing in vain,
but rather in all cases it is for the sake of something, tell me how
nature, who keeps an eye to the future, could produce an organ which
[the embryo] is not yet able to use, [at a time when] nature is enough
53,15K for creation, and the mother's soul is also adequately contributing to
13.6.10 the building of the embryo,[182] but the self-moving soul that acti-
13.7.1 vates[183] [it] is not in use with respect to the works of nature. (7) But
even this, of course, is an ignorant mistake of those who are not able
to understand how the soul is present to the body and again how it
is absent, and that its presence and absence is not spatial; rather it
53,20K is in terms of suitability and harmony that the soul is in [the body]
or in any case present [to it], namely by forming a harmony [with it],
or that it is absent, namely by being incongruent. But that [*viz.* the
13.7.5 embryo] is not yet in a state of harmony as determined by the form
principles of [its] nature. How, then, can [soul] be present to what is
not yet harmonious, when there is no other way for it to be present?
For even if one were to concede that it is present spatially, neverthe-
less the presence that is conducive to the creation of animal life does
53,25K not arise spatially but according to the harmony between the instru-
mental body and the user of that instrument, and when the
instrument is not yet able to form a harmony on account of its being
13.7.10 incomplete, its future user is absent, even if someone should concede
that it is present in another way.[184]

14.1.1 **14.** (1) Yes, they say, but just as the seed has the form principle for
teeth which [the offspring] develops after its delivery, and similarly
53,30K for beards and seed and *menses*, so too are there form principles of
impulse, representation and sensation in [the seed], though their
14.2.1 development [takes place] only after birth. (2) It is obvious that those
54,1K who say these things do not produce anything compelling[185] based on
clear evidence but rather conjectures and probabilities based on their
belief that unless [these form principles] were present in [the seed],
[these features] would not subsequently arise. But their ambition
blinds them to the fact that they are making the soul seminal and
proclaiming the vegetative [power] better than the self-moving
54,5K/14.3.1 soul.[186] (3) But[187] these are the ignorant views of the Stoics who have
turned things upside-down and dared to generate the better from the
worse:[188] they grant being and substance to all things from matter,
and they make nature the offspring of tenor,[189] and the soul respon-
14.3.5 sible for sensation and impulse the offspring of nature, and again the
rational [soul] the offspring of these, and intellect the offspring of the
reasoning [soul]. While they generate everything from the bottom up
through different kinds of and accumulations of motions, one ought
54,10K to proceed from the top down and advance from the better to the
lesser, because every generator is by its own substance naturally
disposed to generate something worse than itself and not something

better.[190] And for this reason the vegetative [power] in us generated something worse than itself, the seed, since it lacks actual movement. As a supplement, it receives the movement from the nature in the mother and from its environment,[191] since in all things the actual precedes the potential. (4) But if there are principles of representation and impulse in the seed and if nature is what then advances these form principles to actuality, then the air that meets [the newborn] at the moment of delivery will, as Chrysippus thinks,[192] be the nature that, having been put in motion, transformed [nature] into soul.[193] And no account could be more materialist (*apsukhoteros*) than this one, because it is godless and dares to produce the better from the worse. But if the soul is not an offshoot but is [nevertheless] retained when it is present in the body, this kind of dramatic account of [the soul's] 'retention' is entirely fictitious and is not the work those who are knowledgeable in these matters.[194] But [such an account] is the work of those men who do not know that the soul is not retained as if by a hand or a chain, or as if by a cage.[195] For its retention is in no way corporeal, but only owing to suitability, just as fire is not retained by a chain or hand, but only owing to the suitability of the matter.

15. (1) If anyone is not willing to direct his foolishness to other pursuits, let him try to show that this thing which is attached and clings to the womb, and which is nourished through the umbilical cord as if through a root, and whose organs of sensation and inner parts are shut and incomplete (for all of its openings are closed off, since [it is immersed] in a fluid in which it will not be able to survive when it subsequently becomes an animal) – [let him show that this thing] is suitable to perceive and have impulse and representation. (2) For I won't even mention that the soul, whose substance lies in cognitive motion, invariably grants, if it is present, this cognitivity to that to which it is present, lest someone think that cases of comas and spells of heavy sleep provide a counterexample,[196] not knowing that even then animals are engaging in representation and simply do not remember these representations, whereas embryos are unable to experience representation since they are not even active with respect to sensation, it being from this that the impressions that go into representations are received.[197] (3) But if someone says that the soul considered by itself does have representation and desire even when it is in the seed but that *qua* belonging to an animal and the composite the soul no longer [has these powers] because an animal is not constituted of a soul and a seed but of a soul and an instrumental body, this man would be saying nothing other than that soul is not yet present to the embryo, to which the soul has not yet imparted its intrinsic activities and which, accordingly, does not have the impulses and representations that occur through soul. (4) Plato and everyone else who admits that the soul [exists] by itself even without

this animal body and who does not give up on there being some
sensations and desires that concern it, say that it sees as a whole so
that[198] it desires and adapts itself[199] as a whole, not through distinct
15.4.5 parts and apertures as when it is in the animal body.[200] When this
animal body, then, is still incomplete, the soul is able to have its
55,20K intrinsic sensations, but how could the sensations that occur with the
15.5.1 complete body be produced [in the embryo]?[201] (5) And they do not
realize that although the deficiencies and excesses of the body create
no obstacle for the vegetative [power] – rather even when there come
to be cuts and amputations and excesses of flesh its natural work is
not obstructed very much – it is not similarly the case that organs
make no difference with respect to psychic powers. At least when a
55,25K/15.5.5 cataract consisting of a little fluid occurs, the power of sight is
obstructed and gone, and when a little bile infects the meninges, the
power of representation is changed,[202] and for the other [organs]
something minor becomes a big obstacle for the psychic movements.
And the embryo as a whole is like clay that is not solid at all and can
be compared to water.[203]

55,30K/16.1.1 **16.** (1) Further, let the seed have a soul – if someone wishes to call
the vegetative power a soul – and let it have this soul from the father
either separated off from him or generated by him, and from the
mother let it subsequently have either nourishment or a power as
56,1K well or even a pneumatic union[204] with the mother's soul responsible
16.1.5 for nourishment and growth – for even Hippocrates seems to call the
vegetative power a 'soul' in accordance with ancient usage[205] – how-
ever, it is impossible to show that the self-moving soul exists in the
seed or even in the embryo because that soul is not even needed, since
56,5K/16.2.1 the seed's own nature has been obtained for this task. (2) For produc-
tive excellence[206] derives from the irrational power, and there is a
need for a flourishing nature that will[207] in nine months increase
that[208] minute amount of moisture to such and such a mass, and form
it and make it solid without being obstructed. But the [activities] of
one [power of soul] that bring it to a state of flourishing obstruct the
16.2.5 other [power of soul],[209] and just as productive excellence belongs to
nature, so too does virtue belong to the soul.[210] Therefore, if there
56,10K were need of soul [for the production of offspring], then it would be
necessary that those men and women who are going to produce the
best [offspring] practice virtue rather than exercising their bodies, as
is the current practice. If, on the other hand, it is physical exercise
that is conducive to productive excellence of nature, and if the
virtuous man, as I think someone said,[211] is a poor husbandman in
16.2.10 the production of children, then the generation of embryos does not
require the soul to which virtue properly belongs[212] but rather the
soul to which [virtue] does not belong and which preserves the proper
56,15K/16.3.1 excellence.[213] (3) To sum up: If the cognitive soul is responsible for the
production of body and it is by means of this soul that we grow and

are nourished, and if it is what forms and nourishes the body, one
must concede that even the individual soul is necessarily in what has
been separated off [from the father].[214] But since the mother's soul is
sufficient[215] for the seed as regards the embryos' needs – and they 16.3.5
need to be formed and nourished and to grow and to have this kind 56,20K
of life and not a cognitive life – the vegetative soul[216] and the creative
power in the life-giving womb are sufficient. (4) Then why do we allow 16.4.1
the difficulty surrounding the soul's source and the scepticism re-
garding the self-moving soul's coming from an external source to
cause an uproar, when [the self-moving soul] does not come from the
mother, nor is the embryo administered by an ensoulment from the
mother?[217] For what is lacking in the [offspring's] nature, which 56,25K/16.4.5
always needs to be 'infused' from the perceptive power because it is,
as we said,[218] an offspring of the sensitive desire and obedient to
sensation (although it has no part in sensation, just as the irrational
part of soul is brought into order by reason, although it is not able to
engage in rational activity), this comes from the mother. (5) However, 16.5.1
regarding the corporeal and irrational substance, what is lacking in
terms of its being joined to [a captain] at birth is provided and afforded 57,1K
by the universe, as an individual soul is immediately present,[219] the
very soul which comes to be present to the [body] that has been brought
forth at just the right moment,[220] and comes to be in harmony with
the instrumental body that is suited to receive it. And the Chaldeans 16.5.5
say that from eternity there has been a divine and intelligible stream
through the eastern parts of heaven. And this stream both moves and 57,5K
turns the cosmos, and brings to life everything in it by sending them
their own souls. And every degree,[221] when it came to be around this
eastern region, which is a portal of souls and the spiritual inlet of the
universe, is given special powers.[222] [This region] was called 'centre' 16.5.10
and [more specifically] 'horoscope'.[223] And everything that has
emerged from a mother or that has in some other way become suited
for being brought to life depends on this invisible stream of individu- 57,10K
als[224], by drawing its own stream from this stream of soul in order to
become alive – for which reason they also call this eastern centre
'place of life'[225] – at the very moment, namely at birth, when the
embryo is released from its vegetative administration and emerges
into the air. And then they make a determination of the time of the 16.5.15
sowing – not that the life-giving stream enters so soon (for how could
what gave soul for the first time when the baby emerged from labour 57,15K
[have already entered at the sowing]?) – but rather that the stream
came to be in harmony with the baby because the baby was suited to
receive it, but it would not have been suited unless the foundation
had been laid at this very juncture.[226] (6) But I have brought these 16.6.1
[men] up not because I find all their views credible,[227] rather I am
showing the trend common to nearly everyone according to which
ensoulment was believed since antiquity [to take place] *after* the 57,20K

delivery from labour of the fetus. And when it is taken into account
that the universe is in agreement both with itself and with its parts,
16.6.5 one must uphold his belief that just as with musical notes the last
note in the scale preserves the harmony running through all the
notes by being in tune with the middle note, whereas if the last note
is too low or too high, it departs from the duple ratio,[228] in the same
57,25K way, the little body of the embryo, when it is in the belly and being
16.6.10 tuned for soul, prior to receiving the adequate 'mode' (*tropon*)[229] of
harmony towards the soul, does not have this soul, but once it is
tuned it immediately has present the soul that will use it. And as long
as the harmony is lacking, the soul is not present, even if the cosmos
58,1K/16.7.1 is chock-full of souls. (7) For just as when cold or heat or some excess
or deficiency has loosened and slackened the harmony of the eye, the
sensitive power is immediately not even potentially present – though
it is clear that for the seeing man {who is out of tune} in the sense of
16.7.5 {having lost one eye}[230] the harmony is present in the other eye – in
the same manner the soul that will [eventually] use the body that has
58,5K been put in tune is absent from the entire instrumental body as long
as the body is still incomplete and out of tune, even if, as I said,[231] all
16.8.1 things are thoroughly covered with souls. (8) Likewise, we are also
unable to go through two or three souls in a lifetime that preserve
our own reason, even granting that the universe is covered with
souls, on account of the fact that we have been tuned to a *single* soul,
and when this harmony is dissolved, the body does indeed become
58,10K receptive to other souls – those of maggots and worms, for example –
16.9.1 but it is separated[232] from its own harmonious soul.[233] (9) … [234] how
could a soul that is of a nature to be present to it not from the outside
and to use it as a whole once it has already become a natural whole
58,15K – how could such a soul be present to this thing [*viz.* the embryo]?

17.1.1 **17.** (1) If not even these [considerations] persuade you, but you
suspect that embryos do partake of the self-moving soul and not of
the nature responsible for nourishment and growth, nevertheless I
am prepared for this too and I say that the external entrance of souls
58,20K into their bodies has not been shunned by Plato. However, let us
17.1.5 grant that the moment at which this occurs is unclear; nevertheless,
that neither the father nor the mother gives the soul to it will be
argued for just as any other point. For it is clear that if it is not from
17.2.1 the parents, it has entered from outside. (2) But let it be debatable,
if you like, whether [the entry takes place] in the sowing of the seed
or in the shaping [of the body] or at the time when it first moves
58,25K spatially or in its delivery from labour.[235] But persuading the sceptics
that the cognitive soul is neither drawn out of nor divided from the
17.2.5 parents, whether in the manner that homoiomerous parts are or that
anhomoiomerous parts are, and regardless of whether [this is sup-
posed to occur] through a diminution of the suppliers or without a
diminution (as is the case with powers)[236] – can be managed through

the following steps.[237] First, we set down the [arguments] that are
able to {support} those {arguing} on Plato's {behalf}[238] who stray away 58,30K
from the truth either voluntarily or because – don't ask me how – they 59,1K
are ignorant of Plato's opinion. (3) For someone might reasonably be 17.3.1
puzzled why a human being does not come from a human being, in
just the way that wheat comes from wheat and horse from horse[239] –
why, rather, is what comes to be from a human being an *instrument*
of a human being. For the vegetative power is an instrument of a 59,5K
human being, while the human being *qua* animal is a kind of compos- 17.3.5
ite, made up of a body and the reasoning soul. In short, if the seed is
potentially a human being, and if we say that it does not have the
soul potentially, how could the seed potentially be a human being
when it has no share in soul? (4) Then, they will claim to observe two 17.4.1
kinds of generation, the one from like and the other from unlike. For
this fire kindled from that fire [is generated] from a like, but the fire 59,10K
that is generated through the rubbing together of stones or sticks, or
through rays descending onto wood shavings has its cause of ignition 17.4.5
from an unlike. The one, then, from like ... from
the.....................suitable for what came to be suit-
able...................came to be animals. And from mud and 59,15K
putrefactions................. <just> as wind-eggs <come to be> not
from a seed...............strongly settling down together from out- 17.4.10
side.............{nature} compelling like to come from like.[240] (5) Just as 17.5.1
it would be outlandish[241] for one to say that the lighting of the wick
from fire came about from outside, in the same way that the light 59,20K
[generated] through knocking stones together was from the outside,
so too would one be hard-pressed (*biaios*) to convince us that that the
things that come to be from seeds – where the things sending forth
the seeds are ensouled – have their ensoulment proceeding from the 17.5.5
outside in a manner resembling the things that do not come to be
from seeds. For what's the point of humanity's elevated nature, [this
objector] will say, if indeed the process of [human] ensoulment is the 59,25K
same as in the case of worms and maggots and <flies and wind
eg>gs[242] whose ensoulment one *has* to concede comes to be from
outside?[243] (6) A human being generates an animal, i.e. not a mere 17.6.1
body but a body having a soul, regardless of whether someone
wishes to refer to this soul as form-principles or powers or a part
or source of life.[244] But what is characteristic of seeds is the 60,1K
constant progression along a certain course and order,[245] i.e. that
this advances after this, and it is not the case that since after birth
it grows teeth and, as time goes on, a beard and pubes, these sorts of 17.6.5
things are not generated from the seed. Moving in terms of nourish-
ment and growth properly arises from seeds as long as it is in the 60,5K
belly, and the motion of sensation once it emerges [from the womb],
and again rational motion upon reaching the right age, and sub-
sequently intelligible motion, even though no psychic addition has

17.6.10 entered from outside.[246] Rather, everything is fused together, just as
in the organic development of fruits, even if some form principles are
separated off at another time, for example in the grain of wheat the
60,10K form principles of the stalk and the leaf and the root were united
before they were separated,[247] but nevertheless the form principles –
and not only of wheat and {fruit, but also of animals – are united in
17.7.1 the seeds}[248] and subsequently distinguished. (7) For just as with a
freshly sprouted nut, [at first] everything is together in a continuous
fresh green mass, but subsequently its shell and the bony substance
60,15K under the shell are distinguished, with the tiny membrane under the
17.7.5 bony substance and the fruit itself within this membrane as well as
the bony [veins] running through the inside of the fruit [all being
distinguished] at different times, and yet the nut's form-principle in
its entirety is always present in the fruit,[249] so too, they say, are all
the parts fused together in the seed [of a human being], but they are
distinguished out of the fusion and individually fashioned and come
60,20K forth {at different times},[250] always with the part suited to the mo-
17.7.10 ment necessarily distinguished before the others. Therefore, since in
the period of gestation[251] there is not yet any need of impulse and
representation, the form principles of these powers lie inactive in the
whole mass,[252] just as the form principles of the stalk and the leaf [lie
inactive] in the grain of wheat, and the moments when each of the
powers advances are always tuned[253] to its needs.

60,25K/18.1 **18.** That the seed is not the product of just the vegetative soul[254]
in us, but also of the soul responsible for sensation, representation
61,1K and impulse, is made clear by how representation rouses the emis-
sion of seed, by how the whole process is accomplished through
18.5 touching and bodily pleasure, and how the seed can be discharged by
certain sensations. They claim, in any case, that sensation is suffi-
cient, since sometimes it flows from the mere sight of the youthful,
61,5K and since representations in dreams leave other [parts] unmoved,
but separates the seed. Therefore, in this way the seed is not a
product of the vegetative [power] alone, but also of the [power]
18.10 responsible for representation. But if it were a product of the [power]
of nourishment and growth alone, it would [of course] be reasonable
for it to have the power to grow and nourish itself, but there is no way
– if it is a product of this sort of soul *alone* – that it will also acquire
from that soul the power to have representations that is manifest in
61,10K the erotic representations as well as the desires and impulses that
follow upon these representations. How, then, could the seed not be
18.5 an automatic[255] product of the soul responsible for representation
61,13K and impulse? Let us not be surprised ...[256]

Notes

1. The noun *eiskrisis* and the verb *eiskrinesthai* seem to have acquired a technical sense relating to the entrance of the soul into the body, e.g. Porphyry Fr. 268.13 Smith; Philo *de Somn.* 1.31; Alcinous *Didask.* 178,35; Plotinus 4.3.9.8; Iamblichus *de Myst.* 1.8; Michael Psellus (in Appendix 2); See Finamore and Dillon p. 165.

2. Since Porphyry often uses locutions such as '*zên phutikôs*' in contrast to *zôion*, the latter must be translated as animal rather than as living thing.

3. As Kalbfleisch points out (13), we know nothing about this Gaurus. The name is extremely rare.

4. cf. Anonymous Christianus *Hermippus* 61,27 (in Appendix 1).

5. The second potentiality is also known as the first actuality. For Aristotle the inability to exercise perception is contained in the definition of sleep (*PN* 458a28-32; cf. *DA* 417a9-11), but whether Aristotle thought intellectual capacities were merely potentially present in sleep is less clear (see van der Eijk (1994) 320), though at *GA* 735a9-11 he seems to concede that they are less active in sleep. Cf. *AG* 13.3-5.

6. As Philip van der Eijk kindly pointed out to me, *karos* often has a pathological sense of 'coma' or 'unconsciousness' as opposed to mere deep sleep. Cf. 13.3.3 where *karoun* is clearly meant to be much stronger than mere sleep.

7. Kalbfleisch marks a textual problem before *proselthontôn* at 1.3.5-6. In the critical apparatus he suggests reading *<mêpô> proelthontôn*, but Festugière's suggestion seems more likely: *kath' ho proelthontôn enesti* for *kai* † *proelthontôn enesti*.

8. *ho horos tautês* here has the same sense as *ton kairon aphorisai tês eiskriseôs* in 2.2.3.

9. In what follows Porphyry offers four alternative views regarding the moment of ensoulment: (i) with the creation and release of the semen; (ii) when the embryo is first formed (between first 30-42 days of pregnancy); (iii) when the embryo first moves (between first 3 and 4 months of pregnancy); and Porphyry's own position (iv) at birth.

10. 'Retained' translates *kratêthênai*. As Soranus reports (*Gyn.* 1.43) *kratêsis* can be used in a sense synonymous with *sullêpsis*, the more common word for conception.

11. cf. Straton's view of conception, according to which a pneumatic power is required to bind the male seed to the female with the result that the embryo must be a living thing (see Waszink 'Beseelung' 178-9), though it is not clear to me whether this power was thought to come from outside. The Pseudo-Galenic *Whether What is in the Belly is a Living Thing*, which according to Hanson (2008, 104) 'probably' predates the *AG*, also argues for conception as the time at which animal life begins.

12. *AG* 2.2.1-10 = Fr. 36 des Places. Regarding this passage, see Schibli (1990) 113-17.

13. This is probably above all a reference to Longinus. Longinus, who was Porphyry's teacher before Plotinus (Brisson and Patillon 5223ff.), is known to have commented at least on passages of Plato (*ibid.* 5261f.), and Pythagoras, Hesiod and Pherecydes all formed part of 'la culture de Longin'

(*ibid.* 5226). As Proclus reports, Longinus in comments on the prooimium to the *Timaeus* apparently advanced the view that the generated gods (cf. *Tim.* 91A-B) sent the soul into the semen with the result that both the semen and soul would enter the womb together (Proclus *in Tim.* 1.51,9-13 = Longinus Fr. 10d Brisson and Patillon). Proclus goes on to say that Porphyry argued against this view (presumably in the *AG*, Brisson and Patillon 5290n.11), and Proclus himself opposes it as well, siding with Porphyry that the soul enters at birth. Cf. Proclus *in Tim.* 3.322,24-30 and see note 14 below. A passage in Plato's *Leg.* (6, 775D-E; cf. 681B5-6) indicating a close connection between the character of the offspring's soul and the state of mind and body of the progenitor at the time of conception was understood by some to mean that the soul enters through the semen (Tertullian *de An.* §25.9; cf. Iamblichus *Vita Pyth.* §211). Of course, to say that the progenitor's state of mind and body at the time of conception has consequences for the character of the offspring's soul is compatible with saying that the soul enters at birth (hence, Iamblichus hardly contradicts himself; cf. *de An.* §31-2). Substandard semen would produce a substandard body, and by the doctrine of suitability an inferior soul would come to reside in it.

14. *Rep.* 10.621A5. Ameles is described as a river in the world of the afterlife. We are told that everyone (except Er) had to drink from it in order to be reincarnated and that its waters caused one to forget one's experiences in the afterlife as well as one's previous life. Curiously, we are also told that no vessel can contain its water (621A5-6). All of this was apparently reinterpreted to mean that souls descend into the human seed, this seed being the fluid border between the sensible and intelligible world. Presumably, the forgetting of the intelligible region was then also attributed to the soul's being in this matter. Note that Proclus gives a different physical interpretation of Ameles in his commentary. For him it is the moist air surrounding the earth, which circulates like a river, but he also sees it as a reference to the flux of becoming in the sensible world (*in Remp.* 2.348,2ff.). Yet Proclus might have been aware of this interpretation, since he explicitly rejects the view of 'some' that the generated gods (cf. *Tim.* 91A-B) place the soul in the semen with both then being cast into the womb (*in Tim.* 3.322.24-30).

15. cf. *Theogony* 361, 775 ff. Like the Ameles, the Styx was a river in the underworld. Porphyry had written a treatise *On Styx* (see Frr. 372-380 Smith).

16. *AG* 2.2.7-10 = Fr. 124 Kern. Cf. Fr. 25 Kern = DK 1B10 and Fr. 49.26 Kern.

17. *AG* 2.2.4-10 = Fr. 87 Schibli = DK 7B7. Cf. Kirk-Raven-Schofield (1983) 60n.1. As M.L. West (1971) 25f. argued based on this passage, Pherecydes' outflow was likely 'a stream or fountain associated with the souls of the dead ... and with their return to life'. Cf. Schibli (1990) 113-17.

18. The Hippocratic *Nat. Puer.* 18.1ff. (and cf. *Alim.* 42 where the author reports others' views on the time required). For a discussion of these relative times for male and female embryos, see Lonie (1981) 190-4. Porphyry seems to attribute to Hippocrates only the reports regarding the time required for an embryo's form to be articulated, and indeed this is all that *Nat. Puer.* 18 substantiates. Iamblichus, however, apparently understood the *kathaper historei ho Hippokratês* more inclusively so as to include the view that the soul enters at the time of first formation (Iamblichus *de An.* §31 Finamore and Dillon, and Porphyry Fr. 266 Smith. See Festugière (1944-54) 3.224n.1-3, with the agreement of Finamore and Dillon (2002) 163, regarding this passage of the *AG* being the likely source of Iamblichus' report. According to

Galen (*de Sem.* 94,8-11 De Lacy), by contrast, the author of *Nat. Puer.* makes the moment of *full* articulation (i.e. extremities are separated and hair and nails are formed) the point at which the embryo becomes an animal. Galen supplies solid reasons for his inference: 'Hippocrates' says that the movement begins at this point (*Nat. Puer.* 21.1), and he describes the fetus as a 'child' (*paidion*) that 'moves' and 'jerks' (*ibid.* 30.1) and in its search for more nourishment ultimately ruptures the membranes and is born (*ibid.* 30.9). But other Hippocratic treatises are arguably closer to belonging to the previous group that place the soul in the seed, insofar as one can find descriptions of conception in terms of male and female seeds which each contain a soul and combining on account of the uniformity of soul (*Vict.* 1.28). For further discussion of Hippocratic embryology, see Lonie (1981), Hanson (2008) and Jouanna (2008). Perhaps Porphyry rather has Alcinous in mind. Alcinous' statement at *Didask.* §25.6 ('It follows from the proposition that souls are immortal that *they should enter into bodies, following upon the natural processes which form the embryo* [*paremphuomenas tais tôn embruôn diaplastikais phusesi*]', Dillon trans.) is usually understood to mean that the souls enter at birth (e.g. Dillon 156; Waszink 322), but Porphyry might have reasonably and even correctly understood Alcinous to be saying that the soul enters earlier. Cf. Louis' translation in Whittaker and Louis (1990): 'elles s'attàchent aux embryons au cours de leur développement'. See also Plutarch *Fr.* 105 (Sandbach), Michael of Ephesus *in GA* 199.7ff.; Proclus *in Remp.* 2.33,9ff. Cf. Lesky and Waszink 'Embryologie' *RAC* 1239-40.

19. Presumably Porphyry has the movement of the limbs in mind. If, however, one may include the beating of the heart as a movement, then Athenaeus of Attaleia and Galen would also fall into this category, as they said the fetus becomes a living thing at this point (18 days after conception). See Waszink (1954) 179.

20. The Hippocratic *Nat. Puer.* 21.1 (cf. Lonie (1981) *ad loc.*). The botanical language here is striking. Once it has progressed through the vegetative activities, it begins to act as an animal by moving. Modern embryology places the beginning of movement already at twelfth week, even though the mother cannot perceive the movements at this time (see Sadler (2004) 121).

21. The theory described here falls into the first category in which the soul is said to enter into the semen.

22. The womb's sympathetic response is also a desire, cf. *prothumia* at *AG* 2.3.6. That the womb is at it were a living thing that has a desire (*epithumia*) to procreate is a Platonic view (*Tim.* 91B-C). Cf. Plotinus 4.4.28.13-14.

23. *metakinêsanta*. Grammatically the subject is either masculine singular or neuter plural. Even though Porphyry just said that the male and the womb are working together to get the soul, the participle is probably referring only to the male here. He is the one who seizes the soul and with it transforms the semen's nature into soul. In other words, on this theory the offspring's soul enters into the father's body and then exits with the semen. The womb's own eager desire comes into the explanation later. It accounts for why this soul is retained by the womb (2.3.6). See notes 22 and 27.

24. cf. 10.3.7 [= 46,30 K].

25. The genetive *tês psukhês* appears to be objective rather than subjective. Cf. Iamblichus *de An.* §32.2-7, which describes this same position using the same terms (*prothumia, helktikê idiotês, anapnoê*, etc.). As Finamore and Dillon note in their commentary *ad loc.*, this 'has affinities to the doctrine of the Orphics' (165). The Orphics seem to have advanced a doctrine of soul according to which there was a single universal soul that was somehow subjected to division into individual souls through individuals breathing in the universe's air (cf. Iamblichus *de An.* §8 and §25 with Finamore and Dillon *ad loc.*). Cf. *SVF* 2.782-4 and Galen's *dunamis helktikê* of the uterus with respect to the semen (*de Usu Part.* 2.323,9-10 Helmreich).

26. The Greek word *sullêpsis* can mean both conception and seizure (or trapping). Soranus (*Gyn.* 1.43) and Galen (*de Sem.* 66,7-11 De Lacy) give similar though less metaphysically loaded explanations of *sullêpsis*, where it is not the soul that is seized, but the semen. Cf. Michael Psellus *Opuscula psychologica, theologica, daemonologica* 23 (98,16-20 O'Meara), translated in Appendix 2.

27. Porphyry's own view is that the semen does contain a portion of the father's nature and carries this with it into the womb. See Introduction § 2.

28. Following Kroll's suggestion of changing *ara* to *hama* here.

29. Here I follow Festugière by translating *paterôn* at 2.4.7 with 'parents'.

30. *to kuoumenon*. See Introduction §4,

31. For *katatattein* in this sense, cf. Soranus *Gyn.* 1.37.

32. Porphyry is surely speaking casually here and does not mean to deny that breathing also takes place through the mouth. Cf. Aristotle *PN* 473a15-b1. He is presumably restricting breathing to the nose rather in order to exclude fish (cf. Aristotle *PN* 470b22-4).

33. The term here is *diapneisthai*, which is sometimes contrasted with *anapneisthai*, the former referring to arterial breathing through pores and the latter to breathing through the lungs (e.g. Galen *de Plac. Hipp. et Plat.* 530,15ff. De Lacy [= 5.710 Kühn]; see Rocca (2003) 227n.89). In contrast to respiration, such transpiration was something that plants and animals had in common (see Marcus Aurelius *Med.* 6.16.1-2 and Farquharson's note [vol. 2, 543]; cf. *de Plac. Hipp. et Plat.* 376,31-3 De Lacy [= 5.525 Kühn]). I doubt that Porphyry means to say that the taking in of air for plants is a distinct process from taking in moist nourishment (but see Lonie (1981) 219), rather I suspect that the *pneuma* is thought to be contained in the moist nourishment, as Anonymous Christianus says (*Hermippus* 1.8 [= Democritus DK 68B5, ergänzender Text 2]). The Hippocratic *Nat. Puer.* 22 also incorporates *pneuma* into its discussion of the genesis of plants, though strictly speaking it does not say that plants breathe.

34. The pith or medulla refers to the soft interior of a plant (i.e. the xylem) as opposed to the harder outer shell or bark. Sometimes one finds *enteriônê* translated as 'heartwood', but this is misleading insofar as 'heartwood' now refers to the inner wood of a tree that has been hardened and made functionless by age. Porphyry seems to think that the plants breathe in both nourishment and *pneuma* through their piths (see previous note). Cf. Themistius *in DA* 42,7-12.

35. cf. *AG* 10.3.

36. This was in fact a contentious point. 'Hippocrates' and Galen in particular are reported to have been of the view that the embryo does take nourishment through the mouth (cf. Michael Psellus *de Omnifaria Doctrina* 115 = Porphyry Fr. 267 Smith). In general, see Lesky and Waszink, 'Embryologie', *RAC* 1235-6, and more recently Congourdeau (2007) 235-9. Perhaps Porphyry thought that the only grounds for this belief were their experiences of peculiar culinary desires during the pregnancy, which they attributed to the fetuses (cf. Tertullian *de An.* 25.3; Soranus *Gyn.* 1.48-53). If so, then Porphyry's response to this view is contained in his Platonic redescription of the phenomena: such desires stem from the womb rather than from the fetus (see below *AG* 8.4). In fact, Galen *et al.* were right to say that the fetus uses its mouth. Starting in the fifth month, it swallows its own amniotic fluid (see Sadler 138).

37. cf. Soranus *Gyn.* 1.57; Galen *de Usu Part.* (2.354,14-355,8 Helmreich)

38. Following Kalbfleisch's suggestion of reading *têi* for *tais* at 3.4.13 (37,14 K).

39. cf. *AG* 4.2.8 and 13.6: The embryo does not yet have the appetitive soul (only a part of it).

40. In *Tim.* 91A4-B4 Plato describes the seed as having a 'lively desire' (*zôtikê epithumia*). Cf. *Tim.* 73B-D and 86C where the seed is said to be a part of the marrow which was created to be the seat of the soul, and 77bB3-4 where plants are also said to partake of the appetitive part of soul (cited below, *AG* 4.4.2-18).

41. *Tim.* 69C5-D6, 70D7-8, 77B3-6; cf. *Rep.* 558Dff.

42. *Theaet.* 156B-C and *Tim.* 64C7-D3. Cf. *Tim.* 42A3-B1, 77B3-6; *Rep.* 583E5-584A1. See also Plotinus 1.1.1.12-13; Michael Psellus *Opuscula psychologica, theologica, daemonologica* 73,17-24 O'Meara.

43. *Rep.* 439B1. Cf. Plotinus 6.8.2.3-5.

44. That is to say, they attempt to address the problem at hand, which is whether embryos should be categorized as animals or plants, by arguing that they are animals, because saying that they are plants seems absurd to them. But their argument actually amounts to the claim that there are no differences between plants and animals, which is even more absurd.

45. cf. Proclus *in Remp.* 2.12,13ff.

46. Porphyry has *auto* for Plato's *autôi*. There is no significant difference in sense.

47. Porphyry has *einai zôion* for Plato's *zôion einai*. There is no difference in sense.

48. Porphyry has *gar* for Plato's *gar oun*.

49. *Tim.* 76E7-77C5. This translation follows Zeyl (2000), which seems to be the modern consensus, but it should be noted that Galen understood 77B6-C3 very differently (on this see my 'The Secret of Sentient Vegetative Life in Galen'). For him the participles in these lines are being positively attributed to plants rather than being denied of them. For an English translation along these lines, see Archer-Hind (1888). Given Porphyry's interest here in maintaining the division between plant and animal, a translation along the lines of Zeyl's seems appropriate.

50. Kalbfleisch's *kuêthen* for *zôion* in the MS. The MS text would read: '... it has the self-moving soul that a [genuine] *animal* has ...]'.

51. *Tim.* 77A3-5. Here Porphyry has *heteron zôion einai* (with Plato), whereas in the passage (4.4.6-7 = 38,21-2 K) above he had *heteron einai zôion*. The addition of *ta hemera dendra kai ta spermata* forces one to translate these lines slightly differently than above.

52. *Tim.* 77B3-6.

53. Reading *ouk esti* with Kroll for *ouketi* at 4.7.4 (39,22 K).

54. *Tim.* 77A6-B1 = *AG* 4.4.8-9.

55. cf. *Tim.* 67B3.

56. Following Kalbfleisch's suggestion of reading *hôs heterôs legomenês kai plêgais kai pathesi hupokeimenês psukhês* for *hôs heterôs kai plêgais kai pathesi legomenês psukhês* at *AG* 4.8.9-10.

57. Porphyry might have the Stoics in mind here who, being materialists, explained sensation in terms of a physical impression in the soul (cf. *SVF* 1.58, 141, 484; 2.53, 55-6, 59, 96. And see Pohlenz (1948) 1.61 and M. Frede

(1999) 300ff.). His reference to *prospatheia* (see note 58 below), however, does not seem particularly apt for the Stoics.

58. 'When the soul is attached to the body' translates *prospatheias genomenês*. As Festugière points out, *prospatheia* is a quasi-technical term in Porphyry that refers to the soul's attachment to the body (*Sent.* 28.10, 29.9, 32.107; *de Abst.* 1.30.4 and 1.31.5; and *Marc.* 32 [125.5 des Places and Segonds]). See the discussion by Jean Pépin in Brisson (2005) 598.

59. *Tim.* 28A2-3. Porphyry's text differs from 28A1-3 by having *perilêpton* for Plato's *doxaston*. Here Plato seems to contrast the *perilêpton* of the Forms with the *doxaston* of the sensible world. Yet Porphyry's version is vindicated at 52A7 where we find *doxêi met' aisthêseôs perilêpton*. In order to preserve the impassibility of the soul, Neoplatonists distinguished two stages in perception. The first involves the body (in some sense), and in particular the sense organs, being affected by the external objects, while in the second the soul actively makes judgments on the affections suffered by the body. In this way, the soul can be said to remain impassive, and its role in sense perception can be called 'judgment' (*krisis*) or 'opinion' (*doxa*) (cf. Blumenthal (1996) 121ff.). What is at issue here is the subject of the first stage of sensation. An apparent disagreement between Blumenthal and Emilsson concerning the subject of such passive sensation in Plotinus provides a helpful background for interpreting these lines of *AG*. What is the purely passive affection that takes places in the first stage? Is it 'merely physical change', e.g. the eye becomes physically red, so that there would be 'no essential difference between sensory affections *qua* affections and, say, affections of books and stones' (Emilsson (1988) 74)? According to Emilsson (*ibid.* 162n.27), this is Blumenthal's view (though I am not sure this is fair in light of what Blumenthal (1971) 72 says about it being the *zôion* in which the impressions are produced). Or is it, as Emilsson himself argues, a more sophisticated change in a metaphysically more complex body? For him the eye must be understood as a live body and the change is the phenomenal colour experience in which even 'three-week old infants' partake (85). Here Porphyry is perhaps suggesting a middle alternative. Whatever the passive change is, it seems to belong to plants (but presumably not to stones). If so, then both the purely physical understanding and the phenomenal reading of this change would seem to be ruled out. The subject of the change is neither a purely physical body nor a body with a sensitive soul (which is presumably the only kind of body capable of phenomenal colour experience). Whatever view Porphyry is suggesting here, it bears a striking resemblance to Plotinus' remark at *Enn.* 4.9.3 where the *threptikon* part of soul is said to be 'passively perceptive' (*pathêtikôs aisthêtikon*). And cf. *AG* 7.3.3-5 (pleasures and pains are expansions and contractions).

60. There are problems with the text here. Kalbfleisch prints: ... *aphantastous † tas pros tês kat' autês tês doxastikês psukhês sunarmosas tou gnôrismatos kai en tois phutois titheis*. In the critical apparatus he suggests emending thus: ... *aphantastous tês tritês kat' auton tês doxastikês psychês sunarmosas tôi gnôrismati* ... This suggestion encounters the problem of ignoring the distinction Porphyry has been underlining between the third part of soul and the activity of opinion. He also documents Diels' suggested emendation: ... *aphantastous tas <tês> pathêtikês kai au tas tês doxastikês psychês sunharmosas, tou gnôrismatos <ti>...*, which is similar to what Festugière suggests: ...*<kai> tas pros tês kat' auton* [*tês*] *doxastikês psukhês sunharmosas*, though Festugière adds that he does not know what to make of the *tou gnôrismatos*. Both of these suggestions encounter a different problem: this *tas* must have the same referent as the first *tas*, namely *plêgas*, which Porphyry would hardly want to attribute to the opinionative soul (*pace* Deuse (1983) 180n.188). I'm afraid something more drastic might be necessary. I am inclined to try: ... *aphantastous* <*ou*> *tais kriseis tautês tês doxastikês psukhês sunharmosas tês gnôristikês* (cf. 16.3.2 and following Deuse [180n.188]) *kai en tois phutois titheis*, etc., and this is what I translate here. However the text ought to look, the idea seems clear enough: Plato is entitled to say that sensation belongs to plants only because he has carefully distinguished the passive motions of sensation from the active motions of the self-moving soul.

61. Here (4.11.1 = 40,28 K) Porphyry has *paredôken* for Plato's *paradedôken* which Porphyry also has at 4.4.16 (39,3 K). There is no difference in sense.

62. Here (4.10.7 = 40,26 K) Porphyry has *parôsamenôi* for Plato's *apôsamenôi* which Porphyry also has at 4.4.15 (39,2 K). There is no difference in sense.

63. *Tim*. 77B6-C3. See note 49.

64. *Tim*. 77C3-5.

65. By not satisfying their cravings, mothers create a birthmark or perhaps even a birth defect in their offspring (cf. *AG* 8.1.6-7). This seems to go against the Hippocratic *Superf.* 18 where the child is said to receive a birthmark precisely when the mother indulges certain cravings, namely for earth and coal. Soranus *Gyn*. 1.48-53 (esp. 53), who also mentions the desire for coal and earth (as does Aristotle *EN* 1148b28), also advises against satisfying any such cravings, though he allows for some exceptions.

66. Kalbfleisch's numbering jumps from 1 to 3. See the Introduction §3.

67. cf. the discrepancy in the times required for articulation and first motion (*AG* 2.2). See Galen *in Hipp. Epid. 2 comment*. 3.31; Soranus *Gyn*. 1.45. Cf. Congourdeau (2007) 228-32.

68. Following Kalbfleisch's suggestion of reading *sunendidomenês* at 5.3.5 for *sunekdidomenês*.

69. But cf. Soranus' example involving a woman looking at monkeys at conception (*Gyn*. 1.39).

70. cf. *Rep*. 454A1-2.

71. That is, to begin with those discussed in *AG* 5.5 concerning representation.

72. This account of *daimons* projecting their representations into *pneuma* should be compared with *Sentences* 29.8-14 (and see J. Pépin's note in Brisson (2005) 594) and *de Antro Nymph*. 11 where Porphyry gives a very similar account of how our souls project forms onto their pneumatic vehicles after death via their faculties of representation (and cf. Sorabji (2004) 1.224ff.). Hence, Porphyry is not saying that humans cannot at all project representations onto matter in the way *daimons* can, since according to him human souls do precisely this after death. The point is rather that this sort of projection is only possible with pneumatic bodies, and so the formation of embryos cannot be due to the presence of the power of representation in the seed. Cf. Michael Psellus *Peri Haidou* 10-15 Duffy, and see below note on *AG* 15.4.

73. On this principle, central to Neoplatonism, that the cause is greater than the effect, see Lloyd (1976). E.g. Plotinus 3.8.5.24-5; 5.1.7.38-48; 5.2.2.1ff.; 5.5.13.37; Porphyry *Sent*. 13; Iamblichus *de Myst*. 3.20; Proclus *Elem. Theol*. §7; *in Tim*. 3.322,1ff., etc. The Neoplatonists understood Plato to be saying this at *Phil*. 27A5-6 (cf. Proclus *in Tim*. 1.259,26-260,4).

74. The text here is difficult. Kalbfleisch prints *hupobebêke <kata> dunameis kai ousias † axia tôn gegennêkotôn* at 6.2.3-4 but in his *Nachträge* (p. 80) he suggests reading *hupobebêke dunameôs kai ousias axiai tôn gegennêkotôn* (Festugière and Limburg follow this suggestion). Deuse (1983) 174n.168 suggests: *hupobebêke <kata> dunameis kai ousias axia<n> tôn gegennêkotôn*. Following Mynas I read *apo* for *axia* and translate: *hupobebêke <kata> dunameis kai ousias apo tôn gegennêkotôn*. Cf. 6.2.6-7: *hupobebêke men kat' ousian apo tou gennêsantos autên nou*.

75. Reading *amoiros* with MS and Deuse (1983) for Kalbfleisch's *emmoiros*.

76. Kalbfleisch suggests (*Nachträge*, p. 80) reading *agetai* for *legetai* at 6.2.10: '... but it is *led* in accordance with reason' (here Kalbfleisch is followed by Limburg and Deuse). Cf. *agesthai* at 6.3.7 and *dioikeisthai* (which seems to have the same sense) at 6.3.3. But neither of these parallels have *kata*. Thus, I follow Festugière in retaining the *legetai*.

77. Porphyry probably has in mind the sort of farming techniques described in Theophrastus *de Caus. Plant.* 1.15.4: trimming and regulating food, sunlight and wind. Regarding the affections of plants, see *AG* 4.8-10.

78. Porphyry has already denied (*AG* 6.3.4) that the vegetative soul is party to representation. Hence, when he says here that the object is represented jointly, we should not understand him to be saying that the vegetative soul also has a representation. Rather, insofar as the vegetative soul is united and obedient to the soul responsible for representation, we can consider the two together as the representing subject without implying that each power is itself representing. Regarding such interaction between the higher and lower soul as an explanation for non-hereditary features in embryo formation, cf. *SVF* 2.753. On the Stoics see Boys-Stones (2007) 82f.

79. *AG* 4.8-10.

80. The subject here must have *logoi*, and *physis* clearly has *logoi* (see *AG* 13.7).

81. Kalbfleisch (p. 80): *toutou* näml. *tou haimatos*, welches aus *tên exaimatôtheisan* (*trophên*, Z. 20) zu entnehmen ist. Cf. Porphyry *Quest. Homer.* 583.

82. cf. *SVF* 3.119 and Epicurus *Fr.* 410 (Usener). It is telling that both are *materialists*. Porphyry's examples of withering and sprouting suggest that he is thinking of the entire plant as expanding and contracting. See *SVF* 3.378 (= LS 65C), 391 (= LS 65B), 463 (= LS 65D); Galen *de Plac. Hipp. et Plat.* 246,38-248,13 De Lacy (= LS 65K); Alexander *in Top.* 164,15; *Ethic. Prob.* 126,7-16; Plotinus 3.6.3.17; cf. Pohlenz (1948) 1.149.

83. Heliotropes are the obvious example. Plutarch (Fr. 101) also includes roses and violets. Proclus (*de Sacr.* 149,12 ff.) discusses lotuses in this connection (see following note).

84. The so-called 'selenotropes'. Cf. Plutarch Fr. 101.19ff. and Proclus *in Remp.* 2.161,20-2 and *de Sacr.* 148,10ff. (the latter is translated in part and discussed by van den Berg (2001) 20-1). The only concrete example I have been able to find of such a selenotrope is the white rose, which was alternately called *kunosbatos* and *selenotropion* (Dioscorides *de Materia Medica* 4.37 [vol. 1, p. 196.13 Wellmann], and cf. Pfister (1938) 1450, 34-5).

85. Grape-vines, for example.

86. That is to say, the motion results simply from having a *phusis* or nature without requiring sensation or representation. Cf. Proclus' view that hymns can be 'sung' either *noerôs* or *logikôs* or *aisthêtôs* or *phusikôs*, as plants do (*de Sacr.* 148,13-14).

87. cf. *AG* 5.1.7-8 and note *ad loc.* Cf. also Proclus *in Remp.* 2.33,20-34,2 where Proclus credits these cravings (as well as certain aversions) to nature.

88. Kalbfleisch (p. 80) calls attention to Michael of Ephesus' [Philoponus'] use of *AG* 8.2-3 in his *in GA* 33.24-34.5 (Hayduck).

89. *Tim.* 91A1-3. Porphyry has *ton tês xunousias erôta theoi* for Plato's *theoi ton tês sunousias erôta*. There is no difference in sense.

90. *Tim.* 91B5-7. Kalbfleisch gives the reference without marking this as a quotation, but Festugière and Hayduck (see note 88) rightly do so.

91. Reading Plato's *para* for Porphyry's *peri*.

92. On the 'wandering womb' see Longrigg (1998) 194-201.

93. *Tim.* 91B7-D5.

94. These show appetite, spirit (*thumos*) and impulse respectively.

95. *Tim.* 91D1-2 (quoted above in *AG* 8.3).

96. Following Kalbfleisch's suggestion of reading *phutikê <hê> dioikêsis.*

97. *Tim.* 91D1-2. Porphyry replaces Plato's *adiaplasta*, which he himself uses at 8.3.9, with *aplasta*. The *dia* adds a sense of completion, and I have attempted to capture this through the respective translations 'not fully formed' and 'unformed'.

98. Reading *ekthrepsôntai* (middle subjunctive aorist) at 9.2.6 (45,15 K) for *ekthrepsôsin* (active subjunctive aorist). Festugière retains the latter and gives it a passive sense in his translation ('avaient reçu ... croissance'), but nothing in LSJ suggests that this is possible.

99. This is meant to translate the imperfect *ên*.

100. *Tim.* 91D4-5.

101. The phrase 'the things carried in the womb' translates *ta kuoumena*, which I translate everywhere else as 'fetus'. Here the translation 'fetus' would impair the argument, and so I have adapted the translation to fit the flow of Porphyry's thought. See Introduction §4. This does not appear to be a direct quote from any of the dialogues. Kalbfleisch points to *Epin.* 973D2-4: 'From the start, generation is difficult for every living thing. First, we must get through the state of being fetuses'.

102. *gonê* has several possible meanings in Greek, including 'seed', 'offspring' and '(act of) generation'.

103. *Phaedr.* 248D2-4: *tên men pleista idousan eis gonên andros genêsomenou philosophou ê philokalou ê mousikou tinos kai erôtikou* ('the [soul] that has seen the most [will go] into the *gonê* of a man who will become a philosopher or a lover of beauty or someone with a musical and erotic nature').

104. *Tim.* 42E8-43A6. Festugière inexplicably marks this as a quotation.

105. *Phaedr.* 246C2-6.

106. cf. *AG* 2.1 and 15.3. As Kalbfleisch suggests, the other passages might include *Tim.* 87E5-6; *Symp.* 209B4-7; *Alc. 1* 130A9.

107. Reading *<tês>* before *en* as K. suggests.

108. Here Porphyry is departing from Theophrastus' account of grafting (see especially *de Caus. Plant.* 1.6.1-2 and 10; *Hist. Plant.* 2.14.4; Clement of Alexandria, *Stromateis* 6.15.117-8), according to which the rootstock simply supplies nourishment to the scion in the manner that earth supplies nourishment to all plants. While Theophrastus held that the fruit or flowers that were produced from the grafted scion would simply be better exemplars of the scion's nature (better on account of the preformed nourishment they received from the rootstalk), Porphyry claims that the resulting growth is actually a hybrid that exemplifies properties of each of the constituent plants. See Introduction §2 and Wilberding (2008).

109. *Nat. Puer.* 12.6 (7.488,13ff. Littré). For an analogy to plants in this work, see *Nat. Puer.* 22-7 (esp. 22).

110. Translating *karpôn*. Cf. Theophrastus *de Caus. Plant.*. 1.12.10 with the note by Einarson and Link (1976-90) *ad loc*. Porphyry is maintaining the grafting analogy: the seed is analogous to a flowering scion that is grafted on to a rootstalk (womb). The flower and pod that it initially produces are analogous to the *khorion*. Cf. Aristotle *DA* 412b2-3: pods serve to protect the fruit.

111. Following Kalbfleisch's suggestion (p. 80) of reading *apotelousês* for *apotelousin*.

112. Taking the *ho* to refer back to *empêxis* – the membranaceous shell around the seed.

113. cf. *AG* 3.3. Regarding the role of embryonic and/or spermatic breathing in the generation and development of the embryo, see Lonie (1981) 146-56 and cf. Hanson (2008) 100 n. 17.

114. *khorêgeitai* (cf. *AG* 2.3.4 = 35,12 K).

115. This and what follows should be compared to Soranus *Gyn.* 1.57-8.

116. The Greek word here is *omphalos* (umbilical cord). Porphyry is not observing the distinction made today between the connecting stalk and the umbilical cord proper. The former comes into existence with the creation of the chorionic cavity (12-13 days), while the latter begins to be formed in the fourth week (Drews (1993) 68), being completed around the tenth week (Sadler (2004) 136) by at first incorporating and ultimately obliterating the yolk sac and vitelline duct. Whereas the connecting stalk traverses the chorionic cavity and connects the amniotic sac to the chorion, with the generation of the umbilical cord the amnion and the chorion are gradually fused so that umbilical cord traverses the amniotic cavity, connecting the embryo or fetus to the placenta (see Sadler (2004) 129-36, Schulze (2006) 46-7 and 59).

117. That is to say, external to the embryo but within the chorion. There are a total of three membranes that enclose the embryo: the amnion (innermost), the chorion (middle) and the decidua, i.e. decidua basilis and decidua capsularis (outermost), all three of which ultimately fuse together into the single membrane [see Schulze (2006) 45, Sadler (2004) 129]. Kalbfleisch is unsure whether this refers to the amnion or the decidua (77). Soranus (1.57-8) seems to be aware only of the chorionic and the amnionic membranes (he even indicates that some doubted the existence of the amniotic membrane, though he himself does not). This alone provides good reason for concluding that Porphyry has the amnion in mind here, which was linked to the chorion by the connecting stalk or, using Porphyry's terms, by the umbilical cord (see previous note). The amnion, after all, is the membrane *within* the chorion, and Porphyry seems to be describing such a thing. Cf. the Hippocratic *Nat. Puer.* 14.2 and Lonie 1981) *ad loc*.

118. Soranus says the danger is being destroyed by the surrounding fluid (*Gyn.* 1.58).

119. cf. Aristotle *DA* 413a9; Plotinus 1.1.3.21-3; 4.3.17.21-31 and 21.5-21; Iamblichus *de An.* §16 and §33 Finamore and Dillon; Philoponus *in Phys.* 197,5ff. and *de Opif. Mundi* 278,6-13; Simplicius *in Phys.* 268,6ff. Nature is the ship-builder, the self-moving soul (of the father, the mother and the offspring in turn) is the captain, and the offspring's body is the ship.

120. Kalbfleisch indicates that there is a textual problem with *parholou*, which seems to be a hapax (LSJ and Diogenes list only this instance).

121. As Deuse (1983) 178n.178 and Festugière (1944-54) *ad loc.* have emphasized, the subject here is not the seed itself but the *phutikê dunamis* that is in the seed, as is made clear by 6.4.1 [43,5 K].

122. Following Deuse (1983) 187n.209 I insert a *mê* before *krathenta* in 10.5.6 (47,21 K). As Deuse points out, Porphyry is concerned to distinguish the three familiar kinds of mixture: fusion, juxtaposition and blending.

123. For a slightly different version of the text, see Deuse (1983) 187n.209, though the sense remains the same.

124. cf. Plotinus 4.7.8^5.43-4. The former is the view of the Stoics, the latter that of Aristotelians.

125. Kalbfleisch (p. 21) points to a passage from the *Summ. Zêtêm.* preserved by Nemesius (139-141 [= 43,1-11 Morani], see Dörrie (1959) 70) which relates a similar account of the unity of the soul and the body. This has more recently been affirmed by Zambon (2002), who points to both *Summ. Zêtêm.* and *Sent.* (e.g. 27-8). Yet Dörrie expresses serious doubts about whether Porphyry would refer to the upfront and discursively composed *Summ. Zêtêm.* as 'sacred books', which according to Dörrie are marked by veils hiding a deeper sense, examples of such works being *de Antro Nymph.* and *de Styge* (162-5). Dörrie sees a more likely candidate in *de Regr. An.* Beutler (1953) 290, 8-13 is also not convinced by Kalbfleisch's suggestion and points to this reference as the 'decisive difficulty' that rules out the possibility of Porphyry having composed *AG*. Risch (2007) has recently argued persuasively that Porphyry might be employing *hieroi logoi* in a rather broader sense than is usually assumed, pointing to Galen's use of this expression to describe his own work in *de Usu Part.* (1.174,6 Helmreich), and he also rightly emphasizes that Porphyry only expresses his 'readiness' to write such material, which needn't imply that he ever completed this task (271).

126. Following Kalbfleisch's suggestion of reading *mê<keti êi> kata tên* for *mê kat' autên tên* in 10.6.3.

127. These lines are difficult and have been interpreted in quite different ways. The main issue is how to understand *hê de exôthen pros to ergon tês mêtros episkhuei empneousa*. Generally in *AG exôthen* is used of the individual self-moving soul (as opposed to the *phutikê* soul) since it comes from outside. Hence, Festugière takes *hê de exôthen* to refer to the offspring's individual self-moving soul that arrives at birth and contrasts this with 'the parents' (*hoi men pateres*) in the previous line, but this interpretation stumbles over two problems. First, on this interpretation *tês mêtros* must modify *to ergon*, which raises the question of how the offspring's individual soul can help with the mother's task, if it arrives at birth. There are ways of dealing with this problem. For example, we could take *to ergon* to be either the task for nutrition, as Festugière does, or the product (i.e. the offspring's body). The next problem, however, is more stubborn. In the next line where we are told that this soul 'steers the task, just as human souls <steer> the tasks of their own natures'. For then Porphyry would be saying that this human soul steers its nature's task just as human souls steer their natures' tasks, and this sounds platitudinous as best. Presumably for this reason, Kalbfleisch believes that *hê de exôthen* must refer to the World-Soul (see index entry for *empnein*, p. 68) – still in contrast to 'the parents' – which makes better sense of the last line and probably makes the suggestion that this soul supports the mother's work more intelligible, but this interpretation has problems of its own. First and foremost, as remarked above *exôthen* is used consistently of individual self-moving souls and not of the World-Soul. Moreover, throughout *AG* there is no indication that the World-Soul does anything other than serve as match-maker for individual bodies and souls. Kalbfleisch's contention that the mother's soul works 'unter Aufsicht und mit Unterstützung der Weltseele' finds no support in the text (outside of this contentious passage and the opaque remark at 14.3). It is better, then, to take *hoi men pateres* to have the more straightforward meaning of 'fathers' and to have *tês mêtros* modify *hê de exôthen*: the mother's individual soul. Here Porphyry is explaining his claim that the *phutikê* power of the offspring is never bereft of a captain. When it is with the father, the father's individual soul engages it for its task of achieving its form and completing itself (cf. 10.5.2-3), and when it is with the mother, the mother's individual soul not only fortifies it for its task but also directs that task, since the *phutikê* power is obedient to her and her representations and she has the role of the demiurge (cf. 6.1, 8.2 and 14.3). This interpretation is superior because it (1) makes good sense of the final line; (2) preserves the distinction in roles between the mother and father; (3) preserves the convention of using *exôthen* to refer to individual souls; and (4) eliminates the problems surrounding 'the mother's task'. One might object that it is awkward for Porphyry to use *hê de exôthen* to refer to the *mother's* individual soul in this context, but I think he does this with good reason. He is defending his claim that the *phutikê* never lacks a captain from outside by saying that at different stages different captains from outside are present. This is a possible translation of the text as it stands, although given the poor state the text is in, it is possible that a *hê* prior to *tês mêtros* disappeared and should be inserted.

128. That is to say, the descent of the soul into body is not by force.

129. Prometheus' theft of fire from the gods in order to bring it down to humans was seen by Neoplatonists as a metaphor for the soul's descent into the body. Cf. Damascius *in Phaedonem I* 170,7 Westerink and Proclus *in Remp.* 2.53,8-11. Plotinus frequently likens the goings on in the universe to actions by actors or dancers on a stage in order to explain the compatibility of free-will and providence (3.2.15.21ff.; 16.8-10; 17.23ff. 18.7ff.; 4.4.33.7ff.; 34.28-33; 6.7.7.8-16).

130. Moses in *Genesis* 2:7. Porphyry had familiarized himself with Hebrew scripture (*de Antro Nymph.* 10; *de Abst.* 4.14; cf. W. Theiler (1966) 103). Kalbfleisch points out the parallel with Anonymus Christianus, *Hermippus* 65,1ff. (translated in Appendix 1). *Genesis* 2:7, along with *Genesis* 1 and *Exodus* 21:22-3, were often appealed to in Christian and Jewish embryological debates. See Congourdeau (2007).

131. In order to enter the body through these openings.

132. The Orphics had said that soul is breathed in from air (Aristotle *DA* 410b27-30). Philoponus (*in DA* 186,24-34) offers a metaphorical reading of this Orphic doctrine, but Porphyry here suggests that other Platonists had taken it literally. Cf. Michael Psellus *Opuscula psychologica, theologica, daemonologica* 23 (98,16-20 O'Meara, translated in Appendix 2). Cf. *AG* 11.3.12-18. See also Majercik (1989) 38.

133. *Exapsis* can mean both 'kindling' and 'attaching' or 'fastening,' as the following comparison to naphtha makes clear, but its being kindled (*exaptetai*, 11.2.10) and the soul's being fastened to the body are cases of *exapsis*.

134. Translating *di' holou*, cf. *AG* 16.9.2 (58,12 K).

135. The reference to heaven probably is not meant to suggest that the heaven plays any special role in this theory of sight. Rather it is meant as a jab at the theory's implausible nature, that the ocular cone would go so far so fast. Cf. Johannes Actuarius, *de Spiritu Animali* 8.8: 'To think that cones going out from the eyes reach all the way to the very edges of the heavens in a split-second, or that images of those perceptible objects approach one's perception, these views seem theoretically impossible'.

136. This probably refers to theories of visions, such as those of Ptolemy and Galen, which involved rays being emitted from the eyes and rebounding off of the object of perception back to the eye. See Siegel (1970) 36-7 and 90ff.

137. Epicurus has explained sight through *tupoi* or *eidola* (*Letter to Herodotus* 46.6ff.) that flow off from the object of perception to the eye. See Erler (1994) 148-9, and cf. Plotinus' critique of Epicurus's theory at 4.5.2.12-15.

138. Plotinus 4.5.3.20 and see Emilsson (1988) 57-62.

139. Naphtha is an extremely inflammable liquid that was well known for its capacity to catch fire at a distance. Plutarch, for example, writes that naphtha 'is so sensitive to fire that, before flame touches it, it is kindled by the very radiance (*augês*) about the flame and often sets fire also to the intervening air' (*Life of Alexander* 35.1, tr. Perrin). In general, cf. Plutarch *Life of Alexander* 35 and Strabo 17.15. Michael Psellus, drawing on Porphyry, also uses this naphtha as an example of how the soul enters the body in *Opuscula psychologica, theologica, daemonologica*, 98,24-99,2 [O'Meara], translated in Appendix 2). Cf. Hippolytus *Refut.* 7.25.6-7.

140. The text reads *ho naphthas haphtheis haphthenti puri exaptetai*: 'naphtha is kindled by fire when it and fire are touching'. But the touching language suggests immediate contact, which seems to be the exact opposite of Porphyry's intended point (see previous note). Moreover, the datives are awkward since *haptesthai* requires the genitive. Hence, Diels suggested reading: *naphthas <kai mê> haphtheis haphthentos puros exaptetai*: 'naphtha, even when it and fire are not touching, is kindled'. I am inclined to avoid inserting words and to look for a less drastic emendation that achieves roughly the same sense. *haptheis* might have been corrupted from either *ophtheis* (cf. Hippolytus' discussion of naphtha at *Refut.* 7.25.6.4) or *akhtheis* (cf. Strabo's discussion of naphtha at 17.15 where *prosakhtheis* was corrupted into *prosaphtheis*). *Haphthenti* could also be the result of a subsequent corruption inspired by a desire for assimilation to the (corrupt) *haphtheis*, perhaps of *athroon* (which is still found in M. Psellus [see note above] and which is central to Porphyry's view of sympathy [cf. *exaphnês* at 11.3.8 and *di' holou hama* at 11.3.16]). Hence, I read *ho naphthas ophtheis athroon puri exaptetai*: 'naphtha has been seen to be kindled by fire all at once'.

141. The theory of perception by sympathy that Porphyry is advocating here involved denying that the medium (*to metaxu*) had a role to play in perception. See E. Emilsson (1988) 38 and 47-62.

142. cf. Plotinus 4.5.2.55-6; and Porphyry *Sent.* 6. See E. Emilsson (1988) for a discussion of perception and sympathy.

143. I suspect that something is missing here, and that Porphyry does not think that a magnetic stone can attract *karphê*, which usually refers to small pieces of wood. Thus, I add <*ho êlektros*> before *ta karphê*. Cf. Philoponus *Contra Proclum* 274,21-3: '... just as the magnetic stone while remaining at rest spatially moves the iron, and the amber [spatially moves] the wooden chips (*ta karphê*)' And 274,28-275,4: 'For even the magnetic stone always draws the iron that is placed nearby as long as the natural power in it is not exhausted but preserved, and similarly the amber [always draws] the wooden chips (*ta karphê*) and heat is always attracting moisture and cold always condenses what is nearby'. Likewise, [Alexander] *Problemata* 1.41-4: 'Why does the magnetic stone draw only iron, why is the stone made alive by the filings of this [*viz.* of iron], and why does what is called "amber" draw together bran and bits of wood (*ta karphê*), cleaving to them?' (Similarly cf. Alexander *in Metaph.* 16,5-6, *in Meteor.* 220,30, *Problems and Solutions* §23).

144. cf. *Sent.* 27.

145. cf. Aristotle *PN* 471b30-472a1; cf. *DA* 410b27-30.

146. Deuse (1983) 155 and n.90 suggests that *enkosmios* here is synonymous with *ouranios.*

147. On the doctrine of the soul vehicle and astral body, see Dodds (1963) 313-21, Brisson (2005) 593-6 and Deuse (1983) 218ff.

148. *di' holou hama tou hupokeimenou hê empsukhôsis.* Cf. Anonymous Christianus *Hermippus* 64,22-3: *hê de parousia autês akhronos kai diolou hama to hupokeimenon emphukhousa* (translated in Appendix 1).

149. Reading *akhronos* with Festugière for Kalbfleisch's *akhr.i. arkios* has been suggested by Diels.

150. The course of thought here is the same as in Michael Psellus 98,25ff. (translated in Appendix 2): first the sun example, then the chord example.

151. Kalbfleisch seems to be concerned about whether Porphyry is saying this *in propria persona* and suggests that the 'we' might in fact be expressing the opponents' voice (p. 50, critical apparatus to line 9). As Kalbfleisch points out, Iamblichus (*de An.* 15 Finamore and Dillon = Stobaeus 1.48.8 [317,21-318,15 Wachsmuth]) describes Platonists as introducing the intellect at birth – in contradistinction to the Stoics and Aristotelians who make it arrive later. This report fits some remarks from *Tim.* 43D-44A (and cf. *Phaedo* 76D), and Proclus gives his assent that it is already there at birth but remains inactive (*in Tim.* 3.348,6-21). Hence, if Porphyry is saying this *in propria persona*, as he seems to be, he is doing so in opposition to dominant Platonic tradition. The opposition, however, is perhaps not as great as it first appears. Both Proclus and Porphyry agree that the intellect at birth is present without being present (*in Tim.* 3.348,14 and *AG* 12.3.7-8) and contrast this with the kind of accessible presence that arises (for some) later in life. Porphyry simply goes on to underline the significance of this distinction: although intellect is 'present to' the soul at birth, the soul does not 'possess' it because it is not yet suitable for doing so. Cf. *AG* 13.7.

152. Reading *ouk hôs atopon apodexamenos* for *ouk eis atopon anadeixamenos.* As Kalbfleisch recognized, it is awkward to have *anadeixamenos* with *eis* as we find it here. Thus he suggests reading *hôs* for *eis*, but he also notes a problem with *ana-*. Porphyry often mentions intellect coming late or not at all, e.g. *de Abst.* 3.19.3; Fr. 242.12-15 Smith; Fr. 276 Smith; Fr. 297 Smith.

153. Plato *Rep.* 441A9-B1; *Leg.* 653A5-7; *Ep.* 7 341C6; cf. Galen *de Plac. Hipp. et Plat.* 354.30-2 De Lacy; Synesius *On Dreams* 156.5; Philoponus *in Cat.* 50,12-13; Anonymous Christianus *Hermippus* 62,17-19 (translated in Appendix 1). For Aristotle, cf. *DA* 404b6 and 429a7-8; *EN* 1100a1-3; *Phys.* 197b6-8; *Rhet.* 1384b23 and see below note 159. Deuse (1983) 190n.217 refers to a similar statement by Porphyry in Fr. 242.12-17 Smith.

154. For a recent discussion of suitability in Porphyry, see Aubry (2008).

155. cf. W. Theiler (1930) 104.

156. Kalbfleisch marks a problem in his text: *kai † oude osmê tis memph * * polôn aphaireisthai ton noun*, but suggests in the critical apparatus that the sense is roughly that of *kai ou deos mê tis memphêtai to ta paidia aphaireisthai*, which I translate here, though instead of *ta paidia* I read *tôn paidôn* with Kroll.

157. Kalbfleisch and following him Festugière tentatively refer to *Lach.* 188B, but *Leg.* 653A appears to be a more likely reference, where the Athenian says: 'Thus I claim: the first childish sensation of children is pleasure and pain, and it is in these that virtue and vice first come to be present in the soul, but as for wisdom (*phronêsis*) and solid true opinions, these are a blessing (*eutukhes*) even for him who has arrived at old age. In any case a human being is complete when he possesses these goods and all that is contained in them'. In what follows the Athenian repeats a claim from the *Republic* that children do not yet possess reason (*Leg.* 653B4, cf. *Rep.* 441A7-B1).

158. *Phaedr.* 246C2-6 and 248C7ff. That is to say, after they have relinquished their possession of intellect.

159. *GA* 736b27-8. The authenticity of *GA* 736b16-29 has been questioned by some scholars, but Balme defends the passage in general and this claim about intellect coming from outside in particular, by appealing to the authority of Theophrastus (Wimmer Fr. 53b = Fortenbaugh *et al.* 307a = Themistius *in DA* 107,30-108,18). Cf. Iamblichus *de An.* §15 and Alexander *de An.* 81,13-15.

160. The soul has lower rational powers at birth, but it is still incomplete because it does not yet possess the higher rational powers of intellect. See below *AG* 12.5, where Porphyry explains that the reason (but not intellect) is present even in babies. So too Galen *de Foet. Form.* 76,21-78,11 Nickel (= 673-4 Kühn).

161. cf. *de Abst.* 1.21.1; 3.18.5; 3.19.3; 3.26.12.

162. See above note 36.

163. In other words, parents and teachers must help actualize the rational abilities of the young.

164. That is to say, organs can only be counted as existing when they are complete, but gestation is orchestrated in such a way that when this state of completeness is achieved the fetus is suitable to life outside of the womb. Hence, there are strictly speaking no organs of sensation prior to birth.

165. cf. Plotinus 4.5.2.13 and 56-7; 4.6.1.32-3; Aristotle *DA* 419a12-13 and 25-8. I suspect that this is merely a dialectical objection. In any case, it is not clear to me why this would be incompatible with his sympathy-based understanding of perception.

166. Superfetations are cases of a second conception – most often well after the first conception. In the Hippocratic corpus a superfetus was not capable of life, and so the fact that even the superfetuses give off motions discredits the motions of the genuine fetus as indicators of life and soul. On superfetation in ancient medicine, see Lienau (1973) 98f.

167. cf. the Hippocratic *Mul.* 1.48: 'If the chorion remains in the women and if the mouth of womb is not sufficiently wide, the *menses* goes out less than it should, and the belly becomes big and hard, and chills and severe fever will take her, as well as pain throughout the entire body; and in the area of the belly beneath the navel, there comes to be a weighty feeling (*brithos*) in the womb and a turning motion as if there were an embryo (*kai strophê hôs embruou eontos*)'.

168. Moles, which being similar to wind-eggs (cf. Aristotle *HA* 638a10-25) are lumps of flesh that are produced by some mishap in conception. Modern medicine explains the generation of moles in terms of a deficiency in the female egg for which the male seed then compensates to some extent (Sadler (2004) 60). In ancient medicine the culprit was sometimes located in a deficient seed (the Hippocratic *Mul.* 1.71 [vol. 8. 148,24-150,1 Littré]) or a failure of seed to mix properly (Aristotle *HA* 638a22-5) perhaps due to a lack of heat (*GA* 776a2-4). Cf. the discussion of wind-eggs below in note 240. As Aristotle describes it, the experience of carrying a mole was enough like a genuine pregnancy to trick the would-be parents until the time of parturition (*GA* 775b27-32). Cf. Aristotle *HA* book 10; Oribasius *Coll. Med.* 22.6 [vol. 3, 65-9 Bussemaker and Daremberg]; the Hippocratic *Nat. Puer.* 30.3 (with Lonie's notes *ad loc.*); Soranus *Gyn.* 3.36-9 (who explicitly denies that moles move). See Dean-Jones (1994) 161-2 and Mayhew (2004) 45-7.

169. cf. Aristotle *HA* 638a22-5 who describes moles as being neither animals (*zôa*) nor inanimate (*apsukha*).

170. Presumably Porphyry is thinking of the fatally misformed – since he is denying that soul comes in. On teratology in antiquity, see Bien (1997). Cf. Plotinus 2.3.12.9-11; 4.4.38.19-22; Aristotle *GA* 770b9ff.

171. That is, in our present state as adult human beings.

172. cf. Fr. 356.5-6 Smith.

173. cf. *EN* 1102b2-10.

174. Porphyry's concern here is presumably that the bodily foundations for sensation cannot be prepared if the sensitive soul is entirely absent. The *logoi* have to come from somewhere. Cf. Plotinus 6.7.5.1-30 and 4.4.22.1-5.

175. Reading *kata to eidos <kai> êremon monon* for *kata to eidos êremon men ên*. Kalbfleisch suggests reading *monon* for *men ên* in the critical apparatus. Cf. 13.1.8-9: *kata to teleion eidos* [...] *kai monon en hêsukhiai.*

176. Kalbfleisch's text is: *houtô ga(r ...)rei men ho* etc. Mynas (see Introduction §3) had suggested *en trièrei* for (...)*rei*: 'in a trireme', but Kalbfleisch notes that there is not enough space in the manuscript for this to be correct. The (...)*rei men* is in opposition to the *ho de* in 13.2.4.

177. cf. *AG* 1.3.

178. Porphyry here is criticizing those who place too much weight on superficial observation (cf. *AG* 2.2) in determining the moment of complete formation (and hence of the soul's entry into the body). He agrees that complete formation is subject to observation (*theôroumenê*, 13.4.6) and that it determines the moment of the soul's entry, but one needs to look beyond the surface and verify that everything is in working order.

179. Kalbfleisch's text skips from §4 to §6 (as does the *TLG*). §5 should start here. See Introduction §3.

180. *Tim.* 70A7-B3 (spirited part in heart); 44D3-6 and 73B1-D5 (rational part in the head). Plato in fact locates the appetitive part 'in the area between the midriff and the boundary toward the navel' (*Tim.* 70D7-E2, and see 71A7-B1); it is rather Galen who locates it more specifically in the liver and attributes this view to Plato (cf. Tieleman (2003) 153-4). Plotinus follows Galen's lead (4.3.23), as does Porphyry here.

181. This suggests that not even the appetitive soul is fully present yet. The vegetative soul is a part of the appetitive (see *AG* 4.1 = 37,27-8 K).

182. There are two difficulties here. The first concerns the compatibility of saying that nature is *autarkês* in creation and saying that nevertheless, the mother's soul also contributes to creation. From what we have already seen, this *cannot* mean that nature can complete the task of creation without the mother's soul, for nature is always in need of a captain (*AG* 10.4-6). It must rather mean that while nature is the primary or proximate cause, the mother is a secondary or co-cause. In other words, the contrast between *autarkês* and *sunergoun* captures roughly the Stoic distinction between a *sunektika* or *autotelos aitia* and a *sunergon*. For connection between *sunektikon, autoteles* and *autarkês*, see *SVF* 2.346 and 351, and in general cf.

Wilberding (2006) *ad* 2.1.3.1. The second difficulty concerns the sense of *hikanôs*. If nature is *autarkês*, in what sense could the mother's soul act *hikanôs*? The answer seems to be that the mother's soul is sufficient *qua* co-cause. In other words, since nature's being *autarkês* does not rule out its needing outside help in the form of a captain, the mother is sufficient for filling this role.

183. As Festugière points out, this is probably meant to pick up on Aristotle's language for describing the actualization of an object of movement by a mover, e.g. *Phys*. 202a17: the mover is what activates (*energêtikon*) the moved, or brings it to actuality.

184. cf. *AG* 12.3.

185. Following Kalbfleisch's suggestion of reading *ouden propherousi biastikon* for *ouden diapherousi biastikôs* in 14.2.2.

186. Kalbfleisch's text skips from §2 to §4, but he corrects this in the *Nachträge* (p. 80). See the Introduction §3.

187. Kalbfleisch marks a textual problem at the start of 14.3.1. As it stands, the sentence begins with *ara*, which is impossible as *ara* is postpositive. I read *alla* for *ara*, as he suggests in the critical apparatus, though Festugière's suggestion of *hora* is also possible.

188. This discussion should be compared to Plotinus' critique of the Stoics in $4.7.8^{3}$, on which see van Straaten (1975). Cf. also 3.1.8.7-8: 'Soul does not come to be from seeds, as other things do, rather it is itself an initiating cause'. And see Deuse's (1983) 184f. discussion.

189. I have adopted Long and Sedley's translation of Stoic *hexis*, which refers to the state of the pervading *pneuma* that sustains inanimate objects. See Long and Sedley (1989) 289.

190. See above note 73.

191. Here Festugière warns: 'one dare not be more specific'. He suggests that Porphyry might have the new-born in mind, in which case the environment would be the surrounding air. For Plotinus, who makes a similar remark at 2.3.14.29-33, the environment is always a contributing cause and seems to include material, seasonal, dietary and astrological influences.

192. For the doctrine that the cold air at delivery is responsible for changing the *pneuma* from the state of *phusis* to the state of soul. Cf. Plotinus $4.7.8^{3}$ (= *SVF* 2.804); Tertullian *de An*. 25 (= *SVF* 2.805); Plutarch *de Stoicorum Repugnantiis* 1052F and 1053D, *de Primo Frigido* 946C, *de Communibus Notitiis adversus Stoicos* 1084E (= *SVF* 2.806); Porphyry Fr. 249 Smith. For further passages see Waszink (1947) 321 and Babut and Casevitz (2004) 319. Porphyry argues here as follows: Chrysippus says that air is what causes the change to soul, but he also says what causes the change to soul is nature. Hence, he ends up saying that air is nature, which is absurd.

193. There is a point of tension in our sources regarding whether soul is acquired in addition to nature or whether nature rather completely transforms into soul. See Babut and Casevitz (2004) 319-20.

194. Kalbfleisch's text at 14.4.8 (54,22 K) is *oudamou tôn gig...rist...*, though he adds (80) that the second 'g' and 's' are not certain. Partially accepting Festugière's suggestion, I read *oudamou tôn gnôristikôn*.

195. cf. Porphyry *Sent*. 27; *de Phil. ex Orac*. 178.1-2. Theiler detects Ammonian doctrine here, cf. (1966) 107.

196. 'Provide a counterexample' translates *legein ti*, which here has the sense: to say something relevant (*LSJ legô* (B) 3.6), but the relevance in question is surely that they might provide a counterexample insofar as someone in a deep sleep is ensouled and yet not cognitive. Porphyry disagrees, insisting that cognitive work is still going on.

197. cf. Damascius *de Princ.* 1.30 [1.44,7-8 Westerink and Combès].

198. Reading *hôste* as Kalbfleisch suggests for the MS's *hote*.

199. Translating *oikeiousthai*. Cf. Porphyry *Sent.* 29.22-4 Brisson = 19,4-6 (Lambertz).

200. As opposed to its *pneumatic* body. See below note 201.

201. Porphyry's views on the vehicle (*okhêma*) of the soul must be taken into consideration here. The question at issue concerns the self-moving soul's relation to lower powers such as sensation. Does the soul have them only when it is in the animal body (hence the emphasis on *animal* [*zôikôi*] body at 15.4.5 as opposed to *pneumatic* body) or does it also have them when it is separated from it? Iamblichus (*de An.* §§10, 13 and 37) reports different responses to this question, including Porphyry's: these irrational powers do survive, but they are dissolved into the universe from which they came. This is because the soul's vehicle is necessary for such psychic activities, and this vehicle, according to Porphyry, initially survives death but is ultimately dissolved into the universe (Proclus *in Tim.* 3.234,18-26, on which see Dillon (1973) 372-3). After death, when the soul is free of the animal body, it can continue to engage in sense perception and representation using the vehicle *as a whole* (Hermeias presents a similar doctrine, *in Phaedr.* 68,18-23). Since the entire *okhêma* can be used as a medium for sensation and representation, forms can be projected onto it (cf. *AG* 6.1 and note there). Porphyry seems to be suggesting here that during the period of gestation the soul has already collected its vehicle from the universe, giving it access to these powers of perception. Regarding Porphyry's views on the soul's vehicle, see W. Deuse (1983) 218-27 (who argues at length that the irrational soul is not to be *identified* with this vehicle and addresses its relation to sensation and perception); E.R. Dodds (1963) 318-19; J. Pépin's note in Brisson (2005) 593-6. Cf. Philoponus *in DA* 19,31-6 and 239,2-5; Michael Psellus *oratia minora* 29,23-6.

202. The meninges are the membranes surrounding the brain. The notion that bile and other physical substances could affect the mind goes back at least to the Hippocratic *On Sacred Diseases*, cf. van der Eijk (2005) 45-73 and 131-5. Aëtius Amidenus (*Tetrabiblon* 6.8 [esp. 137,3ff. Olivieri]) takes this a step further by specifying the meninges. Plotinus similarly thinks that bile can affect one's representations (4.4.28.35-43; cf. 6.8.3.12-16).

203. The text is partially destroyed: *to d' embruon holon {p}êlôi eoike mêdamôs st{ereôi}, hudat{i} oiôi {te} onti {paraballesthai}*. The idea here is presumably not only that the embryo is not yet formed (so that sense-perception cannot take place), but that in its unformed state it is even resembles the sort of thing (a kind of liquid) that obstructs sense-perception in completely formed bodies.

204. *Sumpnoia*. Plotinus uses this term to refer to the sympathetic unity that obtains between different parts of the universe at 2.3.7.17, where Armstrong translates it 'a single united breath of life'.

205. E.g. the Hippocratic *Vict.* 1.28 [6.500-2 Littré], where the mixing together of the male and female seed in conception is described in terms of souls being mixed together. Hanson (2008) has pointed out that *Vict.* 1.25ff. makes no explicit reference to plants and suggests that Porphyry is working exclusively with *Nat. Puer.*

206. Reading *eugonia* for *eutonia* at 16.2.2 as Kalbfleisch suggests (cf. *eugonia* in *ll.* 9 and 12). Since the *eugonia* of nature is being contrasted with the *aretê* of the soul, a translation such as 'productive excellence' seems desirable.

207. Reading *tês mellousês* with Diels (see Kalbfleisch p. 80) for Kalbfleisch's *têi mellousêi*.

208. Reading *ekeino* for † *kenoi* as Kalbfleisch suggests.

209. In his index (p. 75 *prosienai tini*) Kalbfleisch indicates that he thinks the *prosionta* in 16.2.5 should be taken together with the dative *têi heterâi*. But I think that Porphyry is saying that there are specific activities that bring a part of soul into a flourishing state (*prosionta eis euthenian*). These activities are practicing virtue for the soul and physical exercise for nature. Moreover, the activity that helps the one flourish, obstructs the other. Hence, I take the dative *têi heterâi* as the object of *empodizei*.

210. As Kalbfleisch indicates there is a problem with the text here. He prints *hôs eugonia phuseôs aretê psukhês* † *kai aretê phusis*. Kalbfleisch's suggested emendation seems best: *hôs eugonia phuseôs kai aretê psukhês*. Diels and Festugière both produce suggestions that retain more of the text as printed (Diels: *hôs eugonia phuseôs aretêi* (sc. *empodizei*) *psukhês kai aretê phusei*; Festugière: *hôs eugoniai phuseôs* (sc. *empodizei*) *aretê psukhês, kai aretêi phusis*), but they seem to disrupt the argument.

211. It is not clear to me, nor was it clear to Kalbfleisch ('wer?'), who is meant here, but cf. Plotinus *Enn*. 5.3.8.27-31 and Proclus *in Remp*. 2.91,12-15.

212. Reading *hês idion hê* as Kalbfleisch suggests for *hês* † *idrê*.

213. *Viz*. productive excellence. The argument here is as follows. Nature and the self-moving soul are different kinds of soul that each have their own excellences and hence their own activities that promote their excellences thereby allowing the respective part of soul to flourish. Nature's excellence is producing offspring, and this excellence is promoted by physical exercise. The soul's excellence is virtue, and this is promoted by practicing virtuous acts. Hence, Porphyry argues that if the soul really is responsible for forming the offspring, then by practising virtuous acts one should be able to produce the best offspring, since this is the activity that puts the soul in its top form. But this is empirically false. The virtuous man is in fact worse at siring offspring. Therefore, the self-moving soul is not required for producing offspring.

214. *En apomerismôi*. Cf. 16.1.2: *ekhetô ge tautên para tou patros apomeristheisan ê gennêtheisan*.

215. Kalbfleisch's text has *arkousês* entirely in parentheses, signalling that the text here has been destroyed. *arkousês* is perhaps justified by 13.6.8-10 where the mother is described as *hikanôs sunergousês tôi kataskeuasmati embruou*. The mother's soul is a sufficient co-cause of the creation and formation of the embryo. I am tempted, however, to read *sunergousês* here instead, which would perhaps be better in light of the *autarkês* – *sunergein* distinction in 13.6.8-10. Likewise, *kai hê en zôtikêi mêtrai dêmourgia* should perhaps be: *pros tên en zôtikêi mêtrai dêmourgian* (cf. 13.6.8-9). If both of these changes were made, the translation would be: 'But since the mother's soul contributes to the seed as regards the embryos' needs [...] the vegetative soul is sufficient for creation in the life-giving womb'. See above note 182.

216. *psukhê* here is entirely enclosed in parentheses, signalling that the text has been destroyed.

217. Accepting Festugière's suggestion: *hotan apo tês mêtros mêketi* (*sc.*

hêkêi hê autokinêtos psukhê), mêde to em{bruon} êi kata tên for the MS's *hotan † apo tês mêtros mêketi to en hôi kata tên.*

218. *AG* 6.3.

219. The universe's role is to provide the individual soul to the suitable body when it is complete. Cf. the use of *sunartan* at 10.4.5 and 10.6.6-9.

220. Reading *hêtis an* <*êi*> *k{ata kairo}n psukhê* with Festugière for Kalbfleisch's *hêtis an k......n psukhê.*

221. As Sextus Empiricus relates (*AM* 5.5), the Chaldean astrologers – and I shall ignore here the problems surrounding what it means to call something 'Chaldean' – divided each of the twelve signs of the Zodiac into 30 degrees (*moirai*) which were then in turn further divided into 60 minutes (*lepta*). For Chaldean astrologers it was important to determine the exact degree ascending over the Eastern horizon at the moment of birth. Thus Sextus reports that they would work in tandem, with one situated outside with a clear view of the horizon and the other, armed with a loud bell, in the delivery room (*AM* 5.27 and 68). See F. Boll *et al.* (1966), 61-3 and Deuse (1983) 154f.

222. cf. Sextus Empiricus *AM* 5.20.

223. Just as astrologers divided the rotating Zodiacal circle into the twelve signs (see above note 221), they likewise divided the static circle through which the Zodiacal circle moved into twelve regions (*loci* – sometimes though not entirely correctly called 'houses'). More specifically, there were four points on this static circle that were considered to be of greater astrological significance. These were all generically called 'centres' (*kentra* or *cardines geniturarum*). (1) The first and most important centre was located at the Eastern horizon. This was called the 'horoscope' (*hôroskopos* or *ortus*); (2) the point on the ecliptic in the middle of the visible sky, called the *mesouranêma* (*medium caelum*); (3) the point on the western horizon, called *dusis* (*occasus*) ; and finally (4) the lowest point on the ecliptic located beneath the horizon, called the *antimesouranêma* or *hupogeion* (*imum caelum*). *Hôroskopos* could be used to refer either to (i) this region of the static circle containing the Eastern horizon, or to (ii) the degree of the rotating Zodiacal circle that happens to be located at this region at birth, or to (iii) the entire Zodiacal sign that contains this degree. See F. Boll *et al.* (1966) 61-3 and F. Boll (1903) 87. See also Sextus Empiricus *AM* 5.12-13.

224. Following Festugière's translation of *kath' hen* as 'qui s'écoule pièce à pièce'. The individual souls are queued up in a line, see Fr. 271 Smith and Bouché-Leclercq (1899) 601-2.

225. See F. Boll *et al.* (1966) 61-3; F. Boll (1903) 87; A. Bouché-Leclercq (1899) 280-1. Paulus Alexandrinus describes a very similar doctrine (*Eisagôgika* 50.17-21), which Olympiodorus includes in his commentary (*in Paulum* 57.19-22 – attributed to Heliodorus, but for Olympiodorus' authorship see Westerink (1971)). Cf. Plotinus 4.3.12.22-5 and 13.1-3 where he seems to be suggesting that the allocation of individual souls (the choices made in the Myth of Er) were written in the stars.

226. There was some discussion in antiquity about whether the moment of birth or the moment of conception was more important for casting one's astrological lot. Vitrivius (*de Architectura* 9.6 [232,26 Rose]) reports that the primacy of conception dates back to Achinapolus (or Athenodorus, the MSS vary). See Bouché-Leclercq (1899) 373-83 and especially 276ff. where he discusses how the Chaldeans related the birth back to the conception. In his commentary on Ptolemy's *Tetrabiblon*, Porphyry himself had produced arguments for determining the moment of conception from the moment of birth, see Boll *et al.* (966) 154 and Frommhold (2004).

227. Although Porphyry seems to distance himself from the views of the Chaldeans here, elsewhere he seems to advocate very similar views. In Fr. 271 Smith he offers an interpretation of Plato's Myth of Er according to which souls would choose among the possible lives that were written in the stars and enter the universe through the *hôroskopos* on the Eastern horizon (see Boll *et al.* (1966) 94 and Bouché-Lerclecq (1899) 601-2 – though I have some reservations about the other details of Bouché-Lerclecq's account, e.g. on his view the soul enters at conception, which contradicts the doctrine of *Ad Gaurum*). Here and in his commentary on Ptolemy's *Tetrabiblon* he seems to accept explicitly the astrological doctrines of the Chaldeans (see W. Gundel and H.G. Gundel (1966) 213-15). On the tensions in Porphyry's views on astrology, see also F. Boll (1894), 112ff. together with F. Boll (1903), 7n.2.

228. 'The last note in the scale' and 'the middle note' translate *proslambanomenos* and *mesê* respectively, both technical terms in Greek musical theory. The Doric scale ranges over two octaves, consisting of four tetrachords:

($_1$a' g' f' ($_2$e')$_1$ d' c' h)$_2$ ($_3$a g f ($_4$e)$_3$ d c H)$_4$ A

The first (*huperbolaion*) tetrachord ranges from a' to e' and overlaps the second (*diezeugmenon*) tetrachord which ranges from e' to h. The third or middle (*meson*) tetrachord begins with the middle note (*hê mesê*) a and overlaps with the final (*hupaton*) tetrachord which ends at H. The final note A in the scale is not included in a tetrachord and must be 'added on'; hence, its technical name *proslambanomenos*. The duple ratio (*ho diplasionos logos*) is simply the 2:1 ratio of the octaves, e.g. between the middle a and the final A. See Vetter (1933) 843.12ff. and Gombosi (1939) 2. For the duple ratio, cf. Plutarch *Moralia* 1138E-F.

229. Porphyry appears to be borrowing another technical word from musical theory.

230. The text runs: *dêloi d' horôn ho a......... hôs .... ho}ti thaterôi parên hê harmonia*. Kalbfleisch notes in the critical apparatus that the *hoti* might well be *eti*. The latter part ('the harmony is present in the other one') suggests that Porphyry is briefly dealing with the case of a man who is blind in one eye. The point of bringing this case up would be to show that it is not a counter-example to this thesis that the harmony is an all or nothing affair and simply depends on the suitability of the instrument. Thus, I would reconstruct this line so: *dêloi d' horôn ho a{narmostos hôs heterophthalmos ho}ti thaterôi parên hê harmonia*. Cf. Aristotle *Metaph.* 1023a4 and Asclepius *in Metaph.* 344,29-32.

231. Last line of 16.6.

232. Reading *ekhôristhê* with Diels and Deuse (1983) 188 for *ekhôrê(sen)* at 16.9.1.

233. This is Porphyry's explanation of spontaneous generation. The human soul departs when the body is no longer tuned to it, but at this point parts of the body become tuned to other souls – namely those of the spontaneously generated creatures such as worms and maggots. This should be compared with Plotinus 3.4.6.40-5 where he says that it is the departing – in this case, human – soul that itself provides the souls of the spontaneously generated creatures before it leaves (3.4.6.40-5; cf. 4.3.8.47ff.).

234. The manuscript is damaged here (3 lines) and the sense is difficult if not impossible to reconstruct. Festugière suggests the following: 'If, thus, manufacturing nature had not established that soul, by its power, had to unite with the body completely, one could admit that the soul did not come to the new-born from the outside (but that it grew with the embryo). On the other hand, since the union is complete, and since the soul that must utilize the body cannot do so insofar as the body is not born and fully achieved, how ...'. Note that this suggestion is incompatible with the text as it stands, and that it turns this into an argument for the external entrance of soul. I suspect that Porphyry is making a rather different point here, as he explicitly starts arguing for the external entrance only in the next chapter. The gist of the argument might be that even if one concedes that the soul does not come from an outside source, still it must be present throughout the whole, which requires that the body be complete. Thus, even if we were to deny the external thesis, soul's entrance is not gradual.

235. On these four candidates, cf. 2.2-4.

236. cf. *AG* 17.6.2 and Fr. 253 Smith.

237. Literally: 'this is not a never-ending [task] through the following', where 'not a never-ending [task]' translates *{ouk a}per{anton}*.

238. Kalbfleisch prints *prôta de ta dunamena e{lenkhein t}ous hu... t{ou Pla}tônos .....enous t{ithemen}* etc., and in the critical apparatus he suggests filling the remaining gap with: *hu{per} t{ou Pla}tônos {makhom}enous*. This, which is what Festugière also reads, would be translated: 'First, we establish the [arguments] that could refute those arguing on Plato's behalf'. In what follows, however, we do not find any *refutations*. Rather, we get arguments that might *support* those who are mistaken. For this reason I read *e{pikourein}* for Kalbfleisch's *e{lenkhein}*.

239. cf. Anonymous Christianus *Hermippus* 64,16-17 (translated in Appendix 1).

240. Six lines of text are badly damaged, though it seems possible to reconstruct the sense. The second objection that Porphyry must meet concerns the difference between generation from like substances and generation from unlike substances. The problem is this: when a substance is generated from a like substance, e.g. fire from another fire, the substantive form of the generated fire does not come from outside but rather from the pre-existing fire. It is only in the generation from unlike substances that the substantive form (or soul) has an external source. In the damaged lines, Porphyry offers two examples. Products of spontaneous generation that arise out of mud and putrefied flesh are recognized cases of generation from an unlike substance (e.g. Philoponus *in Phys.* 107,12-24 and 272,23-273.16; Michael of Ephesus *in GA* 4,1ff.; though cf. Asclepius *in Metaph.* 407.10ff.) where the generated living things must have their source of life from an external source. Wind-eggs also offer an example of this. Although there were some who thought that they were caused by copulation involving weak male seed (Plutarch *Moralia* 38E; Aristotle considers and rejects this *HA* 559b20ff., 638a5ff. and *GA* 750b10ff.), the more common view, which Porphyry seems to present here, was that wind-eggs are generated without any male seed at all (Varro *Res Rusticae* 2.19; Virgil *Georgics* 3.274-5; Pliny *Nat. Hist.* 8.166; Corumella *de Re Rustica* 6.27.4-7; Galen *de Usu Part.* 2.303,11-15 Helmreich; Michael of Ephesus *in GA* 59,33-60.3; Lucian *de Sacrificis* §6 [cf. Hesiod *Theogony* 924f.]; Olympiodorus *in Meteor.* 177,22-8). Moreover, the source of impregnation was usually taken to be external, namely the wind (see Waszink (1954) 176; exceptions here include Aristotle [*op. cit.*], Michael of Ephesus [*in GA* 105,1ff.] and Galen [*de Usu Part.* 2.303,11ff. (Helmreich)], who say the mother is herself the cause). Since according to these accounts the wind-egg was a living thing (even for Aristotle to some extent [*GA*

741a18ff.]), wind-eggs present an example of generation of a living thing from an unlike external source. Human generation, by contrast, is an instance of natural generation, which always must produce like from like (see especially Philoponus *in Phys.* 272,29-273,16; also Alexander *in Metaph.* 213,19-21; Themistius *in Phys.* 39,23-24; Proclus *Theol. Plat.* 1.69,1-2; Simplicius *in Phys.* 313,15-16), and therefore – according to these objectors – should not involve any external source. Thus, Porphyry must explain why in human generation the soul nevertheless comes from outside rather than from the parents.

241. The text is here again badly damaged, but Kalbfleisch is able to fill in the gaps based on the similarities with Anonymous Christianus *Hermippus* 63,14-19 (see translation in Appendix 1). The text he prints is: *{hôsper oun ouk atop}os ho tês thruallidos tên ek puros {exapsin} exôthen phaskôn gegonenai, {epeidê ek tês tôn li}thôn s{un}krouseôs exôthen ên, houtôs kai ho* .... He also suggests in the critical apparatus that the text perhaps should be: *{hôsper oun atop}os ho ... exôthen <mê> phaskôn*, which more closely parallels the *Hermippus* passage and gives a similar sense. Festugière follows this latter suggestion. I, however, have reservations about this construal. In *AG* 17.4 Porphyry's hypothetical objectors established a distinction between generation of like by like and generation of unlike by unlike, where the former was illustrated by fire ignited 'from fire' and the latter was illustrated by fire ignited by rubbing stones together. The objection seemed to be going in the direction of saying that only in the latter case is there a truly external cause of ignition (see above note 240). By sticking so close to the *Hermippus* text, Kalbfleisch's construal eliminates this distinction: 'Just as it would be outlandish for one not to say that the lighting of the wick *from fire* came about from outside, *since it arose externally from the knocking together of stones*, so too ...'. I think better sense can be made of the text by filling in the gaps slightly differently: *{hôsper oun atop}os ho tês thruallidos tên ek puros {exapsin} exôthen phaskôn gegonenai, {hôsper hê dia tês tôn li}thôn s{un}krouseôs exôthen ên, houtôs kai ho* .... In this way the distinction between generation of like from like and unlike from unlike is maintained and given an important place in the argument: Just as it would be crazy to say that the lighting of a wick from fire [like from like] is external, in the way that the lighting from the knocking together of stones [unlike from unlike] is external, so too would one be hard-pressed to argue that generation from seed [like from like] is external, in the way that seedless (e.g. spontaneous) generation [unlike from unlike] is external.

242. Reading *epi skô{lêkôn kai eulôn kai muiôn kai ôiôn hupê}nemôn* as Kalbfleisch suggests in the critical apparatus.

243. In other words, human superiority over maggots and gnats should be accounted for through some significant difference in their respective generations. Therefore, since we have good reason to believe that maggots, being spontaneously produced and as such a case of the generation of unlike from unlike, have an external cause, we must assume that human beings – being superior and thus different – do not have an external cause.

244. 'Source of life' translates *genesis* (following Festugière's *source de vie*).

245. cf. Anonymous Christianus *Hermippus* 64,1-2 (translated in Appendix 1).

246. cf. Anonymous Christianus *Hermippus* 64,3-7 (translated in Appendix 1). This last line makes clear that these are not Porphyry's own views. Indeed, this view looks very much like Iamblichus' (see *de An.* §31 Finamore and Dillon), though I suspect that Iamblichus would not say that these powers are *already in the seed*.

247. cf. Anonymous Christianus *Hermippus* 64,11-14 (translated in Appendix 1).

248. Accepting Festugière's conjecture: *ou tou porou kai tôn {karpôn monon, alla kai tôn zôiôn, hênôntai en sper}ma{si} kai husteron diakrinontai.*

249. Starting from the outermost layer of the nut, we might understand Porphyry's description in the following manner: the 'shell' presumably refers to the exocarp together with the mesocarp; 'the bony substance under the shell' to the endocarp; the 'tiny membrane' to the testa; the 'fruit itself' to the actual seed (the edible part of the nut). This seed consists of three parts: radical, plumule and cotyledons (most species of plant seeds have two cotyledons [cf. Aristotle *PN* 468b19-20 and Ross *ad loc.*], though some only have one). The cotyledons, or seed leaves, make up the largest part of the seed, and they can have a veiny structure, just as ordinary leaves do, only their substance is hard. Thus, I am inclined to understand Porphyry's *tôn entos dia tou karpou dierkhomenôn ostôdôn* as referring to these 'bony' veins running through the cotyledons. However we understand this part of the description, it clearly is supposed to illustrate an articulated state that was not originally there. I would like to thank F. Gavin Hardy for his help on this passage. Cf. Pliny's anatomy of nuts in *Nat. Hist.* 15.86-94.

250. Accepting Festugière's conjecture: *kata diaphorous kairous.*

251. The text is damaged, and Kalbfleish fills in: *en têi kuê{sei}*. Porphyry otherwise uses *kuêsis* consistently to refer to birth (see index), so perhaps some other conjecture is necessary here, but the sense has to be 'in the period of *gestation*'.

252. cf. Anonymous Christianus *Hermippus* 63,22-5 (translated in Appendix 1).

253. Reading *sunarmozein* for Kalbfleisch's *sunomartein*, which would seem to require a dative rather than *eis*.

254. Although Kalbfleisch does produce a continuous text for the first 15 lines, the folio is badly damaged and much conjecture is involved. The first line, for example, runs: *hoti de ouk ergon tês en hêmin {phutikês monês psukhês esti to sperma, all}a kai….*

255. Perhaps read … *oikeiota)ton ergon* for Kalbfleisch's … *autopoiê)ton ergon*?

256. At this point the text is too severely damaged to translate. Nevertheless, some words and phrases can be made out, which allow for a certain degree of speculation regarding the sense of the passage. It is in this spirit that I offer the following reconstruction: Porphyry presents an argument for the opponent's view that all of the psychic powers are already contained in the seed and that no soul comes from the outside. The intimate relationship between the higher powers of soul and the emission of seed testifies to this. This much is presumably an argument that they themselves would give (60,25-61,11). Porphyry takes this argument a step further in the direction of absurdity. If the seed is in a way the product of the sensitive soul, we should not be surprised if it is also the product of the soul responsible for memory and rational thought (61,12-15). A botanical analogy is offered. Everything is already contained in the shoot that is spliced from a tree, but it is not immediately present all at once. Rather, it requires time to achieve its state of completion little by little. So too should we understand the rational part to be already present in the seed and not to come from outside (61,15-22; cf. *AG* 11.3). Porphyry now turns to criticize the view presented above: 'Where do they have such a complete misunderstanding from?' (61,22). One possibility is that they worried about there necessarily being an infinite number of souls if it is admitted that souls come from outside. However, nothing prevents the genesis of the soul from being accomplished via transmigrations (*metensômatôseis*) so that the living come from the dead and vice versa (cf. Michael of Ephesus *in SE* 193,12-16; Proclus *in Remp.* 2.91,19-92.3). Hence, the fear that this doctrine will multiply the number of souls to infinity is empty. Yet the theory that makes the rational soul of the same nature as the souls of the animals so that they might exchange generations is too bold (61,23-62,2. Perhaps: <*ho*> *de logos iskhuros hos homoous{ion tên logikên poiei tais tôn} zôiôn hina (k)ai tas geneseis epallatôsi*. Cf. *de Abst.* 1.19.1-2 and cf. Dörrie (1976) 429-30. And where else do they have this idea that the sensitive and appetitive soul originates in the seeds if not from their observation that certain related character traits come from the same place, namely the parents (62,2-4)? But they are wrong to say that such similarities in character indicate that our higher powers of soul are inherited from our parents (cf. Anonymous Christianus *Hermippus* 63,5-8). Rather, these too are ultimately due to the nature inherited from the father and from whatever the mother contributes (cf. *AG* 3.1; 10 *passim*; 16.1 and 3). This nature is sufficient because it creates a particular kind of body that will then affect the character in certain ways, e.g. a bilious body makes one angry (62,5-9; cf. Plotinus 2.3.14.4-6; 4.4.28.35-43). Another proof is added that the soul's genesis is not a case of like coming to be from like but rather something that enters from outside (Perhaps: *{ex}ôthen* <*tekmêrion*> *prostithetai* for K's *ôthen temakhion prostithetai*. Cf. e.g. Porphyry *in Ptol. Harm.* 78,4-5.). If the soul's coming to be out of the vegetative state is a case of like coming to be from like, then even the vegetative soul will be immortal, since likeness plays an important role in the arguments for the immortality of the soul (62,9-21; cf. *Phaedo* 78B4ff.). Reason and nature are not alike. Unlike nature, reason needs to be aware of its goal just as every craftsman needs to know the immediate goal in order to create. 'For neither a weaver nor a cobbler would try to make anything without first knowing what he was going to make.' It is through this goal and not through the craft all by itself that he creates (62,21-4; cf. Plotinus 4.3.10.11-17). Chrysippus' theory of the creation of soul *via* the impact of the air at birth (cf. *AG* 54,15ff.) is likened to saying that flame is already in a torch or coals and just need to be awakened by air (cf. Long and Sedley (1987) 53B). Presumably this is meant to show that Chrysippus' theory is ultimately Anaxagorean in the sense that for him the generation of the soul is a case of like coming from like (62,24-30; cf. Anonymous Christianus *Hermippus* 63,19-22; Theophrastus *de Igne* §§27-9). The rest of *AG* has perished.

Appendix 1
Anonymous Christianus

Hermippus or On Astrology 2.17[1]

A discussion of generation directly follows, since the ancients were at 61,17 a loss regarding whether one should establish conception as the starting point or the advance of the offspring into the light of day. For if one says that its soul comes from outside, then one would reasonably suppose this [*viz.* birth] to be the more appropriate starting point. But if one says it comes from within, this again raises a problem: when should one consider this to take place? For this critical moment does not seem to be completely determined. Indeed, we have been exposed to various views [on the matter] since our youth. One person says that embryos are not animals because they live only in the manner of plants; another says that they are animals on account of the fact that they will live as animals once they go forth to the outside world. I, however, am most of all convinced that what is 62,1 peculiar to animals has to do with impulse, sensation and representation. But I also believe that it is bizarre to think that embryos have not yet achieved this and that consequently they are not animals. For if the power of the soul is lurking there, as it were, and has to stay hidden for a bit, nevertheless after a short time it reveals itself and orchestrates its own perfection. And I would even say that the soul at that time wanted to remain inactive and achieved this by issuing orders to the powers beneath it. But they say that just as one would not think it right to call the seed that still remains with the father an animal, neither would one say that the seed that has come to be in the womb is such a thing, if it does not become this sort of thing [*viz.* an animal] in actuality. And Plato's [view] on this is an obvious one for showing this, as he calls embryos animals that are unformed and invisible due to their smallness. For they say that what is unformed is not yet an animal. And Aristotle, who says that intellect comes to human beings later and not even to *all* human beings absolutely, maintaining rather that it is rare for any soul suitable to combining with intellect to come to be, seems to agree with these views. And in general they liken this [intellect] to the captain (*kubernêtês*) who stands over the rudder and directs the steering after the complete creation of the ship. This, and things like this, is what they say, and it is largely specious.

I, however, would love to ask them how and whence the soul, which they introduce as a *deus ex machina*, has come into the body. For if

63,1 they say that it flies and is carried round through the air, after the manner of a sparrow, to search for some body to enter through the mouth and nostrils, these men will fill the air with an enormous number of souls wandering in different directions and present this as their best material to those who are minded to believe in the transmigration of souls. And if the offspring receive the characters of those who begat them and show these characters still more clearly once they have become mature, then it is surely clear that the powers of the soul are interlaced with the body. Someone might perhaps say that this [*viz.* having a character that resembles its parents] is not unreasonable even if the soul has come in from the outside. For its powers follow upon mixtures of body. Is it, then, preferable to say that a soul interlaced with the body makes the animal exist or that a soul having come from outside does so? For by saying the latter (*houtô*) we will concede that this soul, when it comes down [to the body], rules the body no more than it is ruled by it. And it would be extremely absurd of us all, if we would not even grant that the kindling of a wick from fire comes about from an outside source, when it had in fact made its acquisition externally from the knocking together of stones, to accept that offspring have their ensoulment from an outside source, even though the producers of seed from which they generated are themselves ensouled. And in general these men seem to me not to realize that fire lies hidden in the torch and waits for the blast of air from the outside that will rekindle it and awaken the flame. And suitability is always predetermined with respect to the right moments. Therefore, if the fetus does not yet have any need of impulse or representation, the form-principles of these powers remain inactive in the entire material substance (*onkos*). Nor would it be possible to include everything all at once, since not even the body is made perfect straightaway, rather some things are required only
64,1 after birth and nature supplies these later. In this manner the soul's powers, too, proceed by some orderly plan to make the appropriate activities appear at the right times. And at first the animal moves in terms of nourishment and growth, as befits the seed, then the motion of sensation emerges and a little later rational motion. But all of these are potentially contained in the spermatic germ and none of them came in from outside.

But they have a problem with how all of these things do not appear in it simultaneously. However, if everything did appear simultaneously, each thing would first be disordered and unarticulated, and then the various [parts] would for the most part be coming into conflict with each other. And even in the budding fruit everything is blended together – the form-principles and the powers – and again in wheat the form-principles of the stalk's leaf and of its root are unified. And similarly in animals that are in undeveloped and unarticulated states, nature keeps the appropriate ordinances safe and intact.

Further, wheat begets wheat and horse horse, and I shall add that even a bird is capable of begetting a bird, and yet a human being is supposed not to generate a human being but only a human body? And besides, the begetter had a soul and the offspring ought to have soul from him. It will not make any difference to me if someone wants to call this soul 'form-principles' or 'power.' And it becomes present in a timeless manner, ensouling the underlying thing all at once. And this confinement is fittingly described not as arising by a hand or a bond, but due solely to the suitability of the thing receiving it, just as we observe that fire, too, is retained (*kratoumenon*) by the suitability of the matter. But we have said enough concerning these things.

But if someone should attack us with the work of Moses, saying that the body was formed first and then the Creator later breathed the *pneuma* into it for a soul that gave life, we should first say that it would not be reasonable to do things any other way. For God must not form and gather up the body by means of a soul remaining spatially outside of the body. Further, I do not see how else the starting point [of ensoulment] could have come about, if indeed the soul has to enter in this way, namely only after the dense masses and porous passages of the body have had their appropriate constitutions perfected, nor do I see how [these points] do not seem to strengthen the position that we are stating with these arguments. For if once the body has been formed and achieved its entire appropriate state of actuality (*kai pasan tên oikeian energeian apeilêphotos*), the soul – wherever it is received from – enters, why wouldn't we say the same thing happens in embryos – that immediately once these have been completely constructed in the womb, the soul straightaway illuminates the body and tries to unveil the appropriate powers? And it was surely not my intention to draw out a story of Alcinous[2] and to extend such a lengthy account for any other reason, but there are many absurdities attached to this view, and if one does not completely destroy their foundation, he would seem to be in a feeble position with respect to what follows from them.

Notes

1. 61,17-65,21 Kroll-Viereck. For two recent discussions of the parallels between this text and the *AG* see Dorandi (2008) and Congourdeau (2008).
2. That is, a long and tedious account.

Appendix 2
Michael Psellus[1]

From *On How the Soul of the Body Enters and How it is Separated*[2]

The manner in which souls enter [bodies], as also the manner in which they are released or separated [from bodies] – both present large problems that are difficult to resolve. None of us preserves the memory of the first yoking together of this combination, and the separation of each animal is something that has yet to take place even as it is happening. How, then, could anyone give an account of what one neither remembers nor has experienced? But since no problem is unsolvable to philosophy, rather since even the ineffable can be stated by it and the esoteric laid out, it is from philosophy that we must take the solutions to these standing difficulties.

First, understand that its entrance is not corporeal, and do not think that the soul coming down from above is coupled with the body by necessity, nor that it is breathed into the body by means of a temporally extended motion, nor that it is drawn up or inhaled from below, nor that it is seized from the air; but also do not think that it introduces itself into the body, as if through apertures, through the nose or mouth or through [other] invisible passages. Neither does the revolution of the heavens, which leads like to like, forcibly propel soul into bodies, nor does a celestial stream proceeding from the east drive or push the soul in. For the soul does not become coupled to the body by moving as if from a distance or through an orbit or incrementally. But if I shall make use of an example for your sake, just as the passage from non-being to being is a kind of instantaneous passage without change, and just as once the sun has risen all things and in particular the surrounding air are suddenly illuminated, and just as strings that are harmoniously tuned immediately resonate with the [played] harmony, and just as naphtha is kindled by fire [simultaneously with] the intermediate airs being ignited, in the same way the body is all at once 'kindled' and illuminated by the soul and becomes filled with its cognitive and seminal powers, without the soul being dispersed little by little into the parts of the body – first being established in the heart or liver or brain and then being spread out from there; rather it seizes the entire body all together and illuminates all at once. And the soul's bond to the body are the harmonic form-principles and the substantial natures of forms [supplied] from universal Nature, from an individual nature, from the parental form-principles, from vital principles and acts, the coherence of the form and the suitability of the body being furnished by higher powers. But angels act as supervisors of this union, both assigning soul [to its place] and protecting it. This is the manner of its entrance.

From *When Embryos are Ensouled*[3]

Of the wise men among the Greeks, some say that embryos are ensouled once they are formed, others again say it is once they move in the belly, and still others say it is when they are detached and perfectly formed for delivery (*olisthon*). The more accomplished of these men say that [the embryos] get the rational soul after delivery, and they transform all at once as if lit up or kindled by divine light. While they are in the womb they *are* alive, but in terms of vegetative and wholly irrational life, being rooted by the umbilical cord as by a root and are furnished with nourishment, but they neither are nor are called animals, as they are incomplete in terms of their participation in the more divine soul. For the rational soul does not, like some kind of bird, thrust itself and enter into whatever it happens to meet, rather it is implanted into what is suited to receiving it through an indescribable union. And how could an embryo be brought to life as an animal in the moist chorion, when the mouth and the nostrils are shut, indeed completely closed? And, if some pure air could even get in there, how could they inhale? And one ought not, they say, conjecture from its movements that this is ensouled life. For it does not act by a representative or impulsive power, but rather as the coils of intestines, when they are full, have motion, and as the bark of trees is moved by rupture, and indeed as plants themselves grow and turn around or close up and then open again with their petals, such as the lotus does with the revolution of the sun (*phôstêr*). But once it comes to be outside of the womb, it detaches from the root and forthwith has a share in the rational soul and is a genuine animal.

Notes

1. In addition to these passages, readers interested in Psellus' possible use of *AG* should look at *Philosophica Minora* 2.102,19-103,4 O'Meara as well as three passages of the *De omnifaria doctrina*: chapters 59, 60 and 115 (this last chapter is included in Smith's *Porphyrius Fragmenta* as Fr. 267). For a good recent discussion of parallels between the *AG* and these passages, see Dorandi (2008) and Congourdeau (2008). I would like to thank Dominic O'Meara for his suggestions on some of the details of this translation.
2. *Philosophica Minora* 2.98,8-99,12 O'Meara.
3. *Philosophica Minora* 1.48,13-32 Duffy.

English-Greek Glossary

acquired power: *hexis*
active, be/to act: *energein*
activity (cf. actuality): *energeia*
actuality (cf. activity): *energeia*
to administer/provide for: *dioikein*
administration: *dioikêsis*
affection: *pathêma, pathos*
affected, be: *paskhein*
air: *aêr*
amputation: *aphaeresis*
animal: *zôion*
animal, land: *khersaios*
animal (adj.): *zôikos*
appetite: *epithumia*
appetitive: *epithumêtikos*
artery: *artêria*
to articulate: *diarthroun*
to assimilate: *katatattein*

baby: *brephos*
barren: *akarpos*
beam/ray: *aktis*
beard: *geneion*
belly: *gastêr*
bile: *kholê*
birth: *apokuêsis, kuêsis*
birth, process of giving: *toketos*
blood: *haima*
bowels: *entera*
to breathe/draw breath: *anapnein*
to breathe in: *emphusan*
breathing: *anapnoê*
bud (pre-grafting): *enkentrizomenon, enophthalmizomenon*
bud (post-grafting): *enkentristhen, enophthalmisthen*

captain: *kubernêtês*
cause (cf. reason): *aitia*
to change into blood: *exaimatousthai*
chorion: *khorion*
class/species: *genos*
cognitive: *gnôristikos*
coma: *karos*
coma, to be in a: *karoun*
completion: *to teleion, teleiôsis, teleiotês*
composite: *sunamphoteros*
conception: *sullêpsis*
consciousness: *sunaisthêsis*
to conspire with: *sumpnein*
control centre: *hêgemonikon*
craftsman: *dêmiourgos*
to create: *dêmiourgein*
creation: *dêmiourgia*

delivery: *exodus, ex ôdinôn proödos, prokupsis*
desire: *orexis*
to desire: *oregesthai*
desirous: *orektikos*
to digest: *katergazesthai*
discursive thought: *dianoia*
to be distinguished out: *ekkrinesthai*
domesticated: *hêmeros*

embryo: *embruon*
to emerge: *prokuptein*
to ensoul: *psukhoun*
ensouled: *empsukhos*
ensoulment: *empsukhia, empsukhôsis, psukhôsis*
to enter (into): *eisienai, eiskrinesthai, engignesthai*
entrance: *eiskrisis*
erotic: *aphrodisios*
evidence: *enargeia*
externally/from the outside: *exôthen*

fashioned, be: *anaplattesthai*
father: *patêr*
female: *thêlus*
fertile soil: *aroura*
fertilization/impregnation: *okheia*
fetus: *to kuoumenon*
to form: *diamorphoun, plattein*

form (cf. kind): *eidos, idea*
form principle: *logos*
formation: *eidopoiia, plasis, plasma*
fruit: *karpos*
fruitfully: *gonimôs*

to generate/engender: *gennan*
to give birth: *apotiktein*
to graft: *enkentrizein*
to grow: *auxein, phuein*
growth: *auxêsis, auxê*

harmony: *harmonia, sumphônia, sunarmostia*
harmony, form a: *sumphônein, sunarmozein*
hibernation: *phôleia, phôleuein*
horoscope: *hôroskopos*
human: *anthrôpinos*
human being: *anthrôpos*

ignorant view/mistake: *agnoêma*
image: *eidolon, eikôn, emphasis*
immediately: *euthus*
implanted, be: *emphuesthai*
impression: *tupos, tupôsis*
impulse: *hormê*
to have impulse: *horman*
impulse, capable of: *hormêtikos*
inactive: *anenergêtos*
inactive, be: *hêsukhazein*
incomplete: *atelês, atelestos*
individual: *idios*
infant: *nêpios*
infant, infancy: *nêpiotês*
infant, be an: *nêpiazein*
inhalation: *eisrhoê*
instantaneous: *akhronos, exaiphnês*
instrument: *organon*
intellect: *nous*
intelligible: *noêtos*
irrational: *alogos*
irrational power: *alogia*

to be joined into one: *enarmozein*
to be joined together/harmonious: *harmozesthai*

kind: *eidos*
kindled, be: *exaptesthai*
kindling/lighting: *exapsis*

labour: *ôdis*
life, living: *zôê*
life, life-giving: *zôtikos*
to live: *zên*
to live as an animal: *zôousthai*
live, living thing (incl. plants): *zôn*
liver: *hêpar*
love: *erôs*

maggot: *eulê*
male: *arrên*
manner of life: *diazêsis*
membrane: *humên, khorion*
membrane, tiny: *humenion*
membranaceous: *humenôdês*
midriff: *phrên, mesos*
milk: *gala*
to moisturize: *ardeuein*
mole: *mulos*
moment (right moment): *kairos*
mother: *mêtêr*

nature: *phusis*
nature, natural merger: *sumphusis*
nature, to form a natural unity: *sumphuein*
nausea: *nautia*
non-living: *azôia*
not firm: *apagês*
not fully formed: *adiaplastos*
not in tune, untuned: *asumphônia, asumphônos*
not involving representation: *aphantastos*
not party to: *amoiros*
nourish: *ektrephein, ekthrepsis, trephein*
nourish, nourishment: *threpsis, trophê*
nourish, nutritive (responsible for n.): *threptikos*

obedience: *epipeitheia*
obedient: *epipeithês, katêkoos*
offshoot: *blastêma*
offspring: *gennêma, kuêthen, tiktomenon*
opinion: *doxa*
organ: *organon*
organ, of sensation: *aisthêtêrion*
out of tune/ill-fitted: *anarmostos*

parent: *goneus*
to partake: *metekhein*
partake, partaking: *metokhos*
perceive: *aisthanesthai*
perceptible: *aisthêtos*
pica: *kissa, kissan*
pith: *enteriônê*
plant: *phuton*
pleasure: *hêdonê*
to pluck: *apodrepesthai*
pneumatic union: *sumpnoia*
poorly nourished: *atrophos*
potential(ly): *dunamei*
potentiality: *dunamis*
power: *dunamis*
premature: *aôros*
presence: *parousia*
present, be: *pareinai*
present in, be: *eneinai*
productive excellence: *eugonia*
to push out: *exôthein*
putrefaction: *sepsis*

rational(ly): *logikos, logikôs*
to retain/master: *kratein*
retention: *kratêsis*
reason (cf. cause): *aitia*
reasoning: *logistikos*
to represent: *phantazesthai*
representation: *phantasia*
representation, to have: *phantasiousthai*
representation, responsible for: *phantastikos*
reproductive: *gennêtikos*
responsible : *aition*
ripe: *pepanos*
ripe, become: *pepainein*
root: *rhizê*
root, to take root/be rooted: *rhizoun*
rub off, to make: *apomorgnusthai*

secretion: *ekkrisis*
seed: *sperma*
seed, to turn something into: *spermatopoieisthai*
self-moving: *autokinêtos*
self-willed: *autokratês*
sensation: *aesthêsis*
sensitive : *aisthêtikos*
to separate off: *apokrinein, apomerizesthai*
separated off, be: *apomerismos*
to shape: *diatupoun*
shaping: *diatupôsis*
shell: *empêxis*
ship builder: *naupêgos*
sceptical, be: *apistein*
scepticism: *apistia*
sleep: *hupnos*
to sleep: *katheudein*
solid, to make: *sumpêssein*
to sow: *kataspeirein*
sowing: *katabolê, spora*
spiritual inlet: *eispnoia*
stalk: *miskhos*
state: *diathesis, hexis*
steer: *kubernan*
steer, steering: *kubernêsis*
still-born: *ta tethnêkota*
stroke: *plêgê*
substance: *ousia*
sufficient: *autarkies*
suitability: *epitêdeiotês*
suitable: *epitêdeios*
suited/suitably: *epitêdeiôs*
superfetation: *epikuêma*
sympathy: *prospatheia, sumpathês*

tendril: *helix*
tenor: *hexis*
to transpire: *diapnein*
tree: *dendron*
tube: *auliskos*

umbilical cord: *omphalos*
unformed: *aplastos*
uterus: *hustera*

vegetative: *phutikos*
vein: *phleps*

waste: *akhrêstos*
way of life: *bios*
wind egg: *hupênemios*
without having representation: *aphantastôs*
woman: *gunê*
womb: *mêtra*
work: *ergon*
worm: *skôlêx*

Greek-English Index

References are to the chapters, sections, and line numbers respectively. Note that the first line of a section includes the entire line of Kalbfleisch's text in which that section begins, e.g. the last word of 1.1, ***sunêkein***, is listed as occurring at 1.2.1, since it is part of the line in which the section break occurs. This manner of reference conforms to that of the *TLG*.

anthrôpos, human being, 4.6.7; 12.2.4; 12.4.1; 17.3.2,3,4,6; 17.4.1; 17.5.6; 17.6.1
antikrus, *mê antikrus,* not at all, 3.4.14
anupêkoos, not obeying, 8.2.9-10
aorata, *a. hupo smikrotêtos,* too small to be seen, 8.3.9; 9.2.5; invisible, 16.5.11
aôros, premature, 10.3.11
apagês, not firm, 10.3.12
apantan, to deal with, 4.2.7
apatan, to deceive, 3.5.4
apathês, unaffected, 11.4.4
apaxioun, to shrink from, 4.2.9
apeinai, to be absent, 12.3.8; 13.7.2,4,10; 16.7.6
apeiria, (not translated), 18.32,54
apeithês, unruly, 8.2.9
apekhein, to be separated, 11.4.1
apêkhêma, faint echo, 12.7.8
aperantos, never-ending, 17.2.7
apergazesthai, to bring about, 2.2.7; 8.1.6; 9.5.5; to produce, 11.2.8
aperkhesthai, to depart, 11.3.1
aphaireisthai, to take away, 12.3.9
aphairesis, amputation, 15.5.3
aphantastos , do not involve representation, 4.10.1; 7.2.2; 12.4.10; 12.7.6
aphantastôs, without having representation, 7.3.6
aphikneisthai, to come, 5.1.5; 9.4.5; 12.3.10-11; to reach, 12.3.10; 12.4.1
aphistasthai , to withdraw, 10.4.5
aphixis, *pros aphixin,* in order to achieve, 12.5.4
aphônos, silent, 12.4.9
aphorizein, determine, 2.2.3,4,12; 2.4.1; to discard, 3.4.5; to define, 4.9.8; *aphôrismenos,* distinct, 15.4.5
aphrodisios, erotic, 18.13
aphrosunê, (not translated), 18.25
apienai, to go away, 11.3.4
apistein, to be sceptical, 17.2.7
apistia, scepticism, 16.4.2
apithanos, implausible, 2.2.3
aplastos, unformed, 9.2.5,5-6
apoballein, to release, 16.5.15
apodeiknunai, to demonstrate, 12.3.12
apodidonai, to respond, 2.2.5; to provide, 4.8.8; 10.6.3
apodrepesthai, to pluck, 9.1.3
apogignôskein, to give up on, 15.4.3; *mê a.,* to deny, 16.6.5
apokeisthai, to be laid up in store, 9.3.2
apokleiein, to shut, 15.1.4
apoklêroun, to ordain, 11.3.5
apokoptein, to cut off, 10.6.4
apokrinein, to separate off, 3.1.6-7; 18.8-9
apokuêsis, birth, 2.2.1; 2.3.12; 2.5.3;
apolambanein, to bottle up, 7.2.1; to extract, 7.2.10-11; to receive, 11.1.8; 16.6.9
apoleipein, to admit, 15.4.2
apomerismos, what has been separated off, 16.3.4
apomerizesthai, to separate off, 16.1.2
apomorgnusthai, to make rub off, 6.1.4, 6,2.1
aponemein, to assign, 13.6.3
apophainein, to proclaim, 3.1.4; 14.3.1
apophrattein, to block up, 8.3.6
apopimplanai, to satisfy, 5.1.6-7; to afford, 16.5.3
apopiptein, to fall, 3.4.10
aporein, to be puzzled, 1.1.4; 17.3.1
aporia, perplexity, 1.1.2; difficulty, 8.3.6; 8.4.4; 16.4.2; 18.49,
aposeiein, to shake off, 11.4.4
apospasthai, to be drawn out, 10.3.11; 17.2.4; 18.21
apostêma, distance, 12.6.1
apotelein, to complete, 8.4.1; 9.2.9; 10.3.4
apôthein, to repel, 4.4.15
apotiktein, to give birth, 5.1.5-6
apousia, absence, 13.7.3
apsukhos, have no soul, 12.6.7; materialist, 14.4.4-5
ardeuein, to moisturize, 7.2.9
arête, virtue, 16.2.5, 6, 10
argein, not to use, 1.4.6; 2.1.3,6; 13.6.10
argos, time-consuming, 5.3.3
aristos, best, 16.2.6-7
Aristotelês, Aristotle, 12.2.3, 12.3.10
arkein, to be sufficient, 16.3.4; 18.6

Subject Index

PORPHYRY

On What is in Our Power

Textual Emendations and Conjectures

187F,5 Reading *tote* with Usener and Festugière for *touto*
268F,36: Reading *ethaumasa hêmas* for *ethaumasamen*
268F,59 Reading *ho* for *ê*
268F,87 Reading *proleipein* with Heeren for *problepein*
268F,88-9 Reading *ta te gignomena dia* with MSS for Smith's *<hoion> ta [te] gignomena dia*
268F,101 Reading *lelumenon* with the MSS for Smith's (following Heeren) *lelumenou*
271F,8 Deleting *prôtêi* before *periodôi*
271F,8-9 Reading *hê de deutera en têi deuterai* with the MSS for Smith's and Wachmuth's *hê de deutera <deutera> en têi deuterai*
271F,14 Deleting *kai klêrousthai* before *keklêtai*
271F,29 Reading *ho helêi* for *helou*
271F,72 Reading *tois gegrammenois* for *tous gegrammenois* (typographical error)
271F,81-2 Reading *hekastou zôidiou* for *tou zôidiakou* as suggested by Stephan Heilen
271F,82 Reading *autôi* with the MSS for *agathôi* (Smith and Heeren), as suggested by Stephan Heilen (2010) 82n.92
271F,93 Reading *tou* with Heeren, Boll, Festugière and Deuse for Smith's *to*
271F,120 Reading *autoi* with Wachsmuth for *autous*

Introduction

Unfortunately we do not possess the entirety of Porphyry's *On What is in Our Power*. What we have is a series of fragments preserved by John Stobaeus (fifth century AD) in his *Anthology* (2.8.39-42), totalling only 14 pages in Smith's *Porphyrii philosophi fragmenta* (295-308).[1] These fragments alone make clear that *On What is in Our Power* was either a commentary or an interpretative essay on Plato's myth of Er in the *Republic* (614B2-21D3), which immediately raises the question of how it is related to what has come down to us as his commentary on the *Republic*. Here too there are only a small number of surviving fragments, this time preserved by Proclus in his own commentary on the *Republic* as well as by Macrobius in his *Commentary the Dream of Scipio*. For the following reasons it strikes me as rather sensible to conjecture that these two sets of fragments originally belonged to a single work.

(i) All of the fragments preserved from Porphyry's commentary on the *Republic* have to do with the myth of Er.[2] This has led some scholars to conclude, in my opinion correctly, that the object of this commentary was in fact limited to the myth of Er, just like *On What is in Our Power*.

(ii) Neither Proclus nor Macrobius ever give us a title of the work by Porphyry that they are consulting, so there is no reason to assume that it is not *On What is in Our Power*.[3] Likewise, Stobaeus never refers to a distinct commentary on the *Republic*. Moreover, given Proclus' high opinion of Porphyry as an exegete of the myth of Er (181T), it is highly unlikely, if there had been a second interpretative essay by Porphyry on the myth, that Proclus would not have mentioned it. Yet it is not mentioned in his commentary on the *Republic* or anywhere else in the Proclean corpus.

(iii) The content that Proclus provides seems to overlap with what we find in *On What is in Our Power*:

(a) 187F of the commentary on the *Republic* and 271F of *On What is in Our Power* seem to be dealing the same astrological theses;

(b) 186F of the commentary on the *Republic* seems to be very relevant to the cosmological account of the soul's voyage through the heavens as it chooses its next life (see below);

(c) Both the commentary on the *Republic* (182F-183F) and *On*

What is in Our Power (see note *ad* 271F,100-2) seem to be concerned to refute Colotes' objections to the myth of Er.

(iv) It is very likely that Proclus' own interpretation of the myth in terms of two different lots was influenced by Porphyry's suggestion of two lives, which is only preserved in the fragments of *On What is in Our Power*.

(v) The opening fragment that Stobaeus gives us (268F) makes clear that a discussion of other topics related to the myth has preceded the ensuing discussion.

Of course, these considerations are not sufficient to establish once and for all the unity of the two works, but I think they do make this thesis more likely than the alternative, namely that there are two interpretative works on the same stretch of text with Proclus only knowing about the one and Stobaeus only knowing about the other. As I see it, the unity question is important primarily because of the pivotal role that 186F can play in helping us understand 268F-71F, as shown below, though of course one might insist on the relevance of 186F to *On What is in Our Power* even while denying that this treatise and the *Republic* commentary originally formed a single work. Despite the above considerations, in my translation below I have decided to follow Smith's sectioning and numbering of the fragments to avoid unnecessary confusion.

The cosmography of choice

The most striking feature of Porphyry's commentary on Plato's myth of Er is his introduction of *two* lives and choices that each soul makes, the first concerning one's species (and in the case of human beings, one's sex), and the second concerning more particular features (see below). I discuss some of the philosophical and other exegetical issues related to this innovation elsewhere,[4] and so here I offer only a short examination of the questions pertaining to *when* and *where* souls are supposed to make these two choices, since these questions bear directly on the translation. Although some of the astrological details of these fragments are difficult to pin down,[5] I believe the general lines of Porphyry's view can be worked out.

There are some genuine difficulties involved in determining just when and where the souls in Porphyry's exegesis of the myth of Er are supposed to make their two choices. Given the fragmentary nature of the treatise, it might be helpful to begin with a brief summary of how Proclus deals with the relevant issues in his massive commentary on the myth of Er (*in Remp.* 2.85,3-359,11). After all, Proclus singles out Porphyry as his Platonic predecessor who was 'a perfect interpreter in particular of all the hidden material in the myth' (181T), so we might expect at least some degree of agreement between Porphyry and Proclus. This expectation is perhaps increased by the fact that Proclus subsequently refers to Porphyry by name only eight times in the entire commentary, despite his high praise of his predecessor's interpretation, which suggests that Proclus might be taking over some of Porphyry's ideas surreptitiously. There are three issues in all that deserve our attention here: (i) the status of the souls making these choices, (ii) where the choices are made, and (iii) what Plato means by 'lots' and 'choices'.

The status of the souls. Proclus emphasizes several times throughout his commentary that Plato is *not* describing souls that are coming down into the sensible world from the intelligible world. He points to relevant features of the myth as well as to other parts of the Platonic corpus to back up his view. First, the choice in question follows the thousand years of subterranean punishments and celestial rewards (615A2-9), and it is obvious that the souls being punished were not in the intelligible region. Nor may one interpret the celestial rewards to refer to a journey to the intelligible region. After all, the duration of a thousand years wouldn't make sense in that case, and more importantly Plato says explicitly that at least some of the souls coming down from the heavens had *not* lived philosophically (*Rep.* 619C8). But a philosophical life would seem to be a prerequisite for succeeding to the intelligible region after death. Indeed, as Proclus reminds us, according to the *Phaedrus* (249Bff.) it is only after completing three philosophical lives that one regains one's intelligible wings (*in Remp.* 2.161,3-8; 2.300,10-2; 2.330,18-331,1; etc.). Moreover, Plato makes a point of saying that the recently rewarded souls tend to make rather bad choices, which is not what one would expect of truly purified souls coming from the intelligible region (*Rep.* 620A2-3). So we are not dealing with souls of philosophers descending again into the sensible world. Nor are we dealing with new souls that are descending for the first time. This is a situation that Plato describes in the *Timaeus* (41E2ff.; cf. Proclus *in Remp.* 2.255,8ff.) and, according to Proclus (*in Remp.* 2.185,23ff.), in the *Phaedrus* (248Dff.), but the myth of Er is clearly describing souls that had already previously been part of the sensible world. It remains to say that we are dealing with souls that are simply 'passing from one birth to another' (2.186,2-3).

The location of the choice. The resolution of the above issue would seem to have consequences for this question concerning the location of their choice. For whereas one might think it reasonable to locate these choices at the outer periphery of the universe if it is souls exiting the intelligible universe that are choosing (though see below), since the myth is dealing with a different category of soul the location-question is still very much open. Proclus concludes that this location is the aether situated at the top of the sublunary region just beneath the lunar sphere. He has two main reasons for this view. First, the souls are said to convene in a place that is between the heavens and the earth, since some souls are said to come down from the one and others come up from the other, and this would fit the aetherial region (*in Remp.* 2.131,14ff.). Second and more importantly, we are told that the souls journey from there to a place of light at which the lots will be distributed and the choices made, and this journey is said to last twelve days. Proclus argues that by measuring this journey in days Plato is making clear that it cannot be taking place above the lunar sphere, since day and night are not experienced up there (*in Remp.* 2.189,8ff.). Thus, the journey must end in the highest reaches of the sublunary aether (*in Remp.* 2.185,23-190,27).

The lots and choices. Proclus has a very interesting view on the lots. He says that there are *two* lots for each soul. The first lot (lot_1) concerns the order of choice. As Plato says, these $lots_1$ are distributed by 'some prophet', which Proclus takes to refer to an intellect in the aetherial region (*in Remp.* 2.254,4-256,21) that distributes the $lots_1$ in accordance with the merits of the individual souls. He says this distribution proceeds 'according to the life that is present in them' (*kata tên zôên tên parousan* 2.254,16), which Festugière is surely right to interpret in terms of their past life which is still present in them in traces (note *ad loc.*), and Proclus seems to think that the prophet's role is only necessary with respect to those souls that are *sunklêroi*, i.e. those souls that approach the choice collectively as we see happening in the myth of Er. (Being *sunklêros* seems just to be a function of the time of death, since the time of punishment and reward is fixed). Nevertheless, Proclus also says that these $lots_1$ are proposed to each soul by the revolutions of the universe (*in Remp.* 2.263,17-18). He appears to make these two claims compatible in the following manner. When a group of souls collectively rises to some point on the outer periphery of the aether beneath the moon, the 'prophet' puts them in a certain order to choose from the selection of lives presented by the rotation of the universe. The idea seems to be that the entire group is assigned a section of the zodiac, with each degree representing a kind of life. Proclus' examples (a kingly life, an erotic life, a combative life – 2.263,22) gives some idea of how he understands the kinds of life in question, but he also warns that there are more kinds of life than we could possibly even conceive of (2.268,10). Thus, the first to choose has the largest selection, with the selection becoming increasing smaller for each subsequent soul, though there are more lives to choose from than souls so that even the last soul has a choice (*in Remp.* 2.265,15-23; 2.273,6-274,21; 2.280,14-282,17). Once this choice is made, a *second* lot (lot_2) is revealed and added again by the universe. This lot_2 fills out many of the details of one's life that are not explicitly included in the primary choice. Proclus' examples here include: type of parents, wealth and poverty, sickness and health, fortunes and misfortunes (2.263,25-9; 2.264,5-17; 2.265,23-266,1; 2.266,11-16). Although he refers to this lot_2 as 'luck' (2.266,13), he makes very clear that this lot_2 is also a matter of 'justice', being allocated to the souls on the basis of what they deserve, and he explains this using the example of the soul that is said to receive the first lot (619B7ff.). Plato tells us that this soul belongs to the group that came down from their rewards in heaven, having lived a good but non-philosophical life, and that this soul nevertheless chooses a tyrannical life that is revealed to involve eating his own children. This is just, Proclus says, because the soul was not really virtuous – Plato describes the soul as choosing out of greed and without reflection (*Rep.* 619B8-C1) – and so even if it did not *act* on its base inclinations during its past life on account of the laws in the city, these inclinations were present and merit punishment (*in Remp.* 2.294,4-295,27).

There is of course much else in Proclus' interpretation of the myth of Er that warrants careful consideration, but this narrow selection of points should help us approach Porphyry's views on the procedures involved in choosing a life. Let us examine his views on each of these issues in turn.

The status of the souls. Proclus presents a strong textual argument for understanding the souls in question not to be descending from the intelligible region but rather to be souls that are already in the universe. Do we have any evidence that Porphyry agreed or disagreed with Proclus on this issue? Porphyry repeatedly refers to the souls as being 'outside' (*hai exô psukhâi* 268F,4 and 82-3; 271F,19), but does this mean outside of their terrestrial bodies or outside of the entire universe? The references to falling 'into their bodies' (*eis ta sômata* 268F,12-13 and 270F,4 and 10; and cf. *pros tina tôn têide biôn* at 271F,18-19) suggests that Porphyry only means the former, but Deuse (158-9) has suggested that Porphyry is primarily concerned here with souls descending from the intelligible region, and that only these souls are in a position to make the first choice. There are several passages that might be seen to support this suggestion. First and foremost is the remark in 271F,66-7 about the souls that 'appear to charge *into the cosmos* at the same degree of the zodiac', which would seem to imply that the souls were originally outside of the universe, and Porphyry's description of the souls 'moving to the ascendant (*hôroskopos*)' might be taken as additional evidence that these souls are moving from outside of the universe to the outer edge of the universe (but see below on the location of choice).[6] Beyond this one could point to Porphyry's remarks in *De antro*, where he appears to endorse Numenius' interpretation of Homer's cave and the parallels he draws to Plato's myth of Er, according to which the *stomia* in the myth are understood to refer to the two gates by which souls move between the sensible and intelligible regions (*De antro* §22 and §29). Nevertheless, there seem to be serious problems with Deuse's interpretation. First, although Porphyry seems to support Numenius' interpretation in *De antro,* Proclus provides a rather comprehensive summary of Numenius' interpretation of the myth including the points about the *stomia* without giving any indication that Porphyry was in agreement with these ideas (2.128,26-132,19), which one might fairly expect if Porphyry advocated this interpretation in his own commentary on the myth.[7] Plus, a *fragmentum incertum* drawn from Macrobius' *Commentary on the Dream of Scipio* and attributed to Porphyry by J. Pépin recasts the Numenian interpretation of the *stomia* in terms of portals between the sublunary and superlunary regions.[8] Secondly, we cannot be sure how strictly we are to understand Porphyry's remark about souls 'appearing to charge into the *kosmos* at the same degree of the zodiac' in 271F,66-7. Moreover, Deuse's thesis that the first choice of species and sex belongs exclusively to souls descending from the intelligible region flies in the face of the details of the text of the myth of Er, according to which none of the souls that chose animal lives would have been candidates for ascending into the intelligible region in the first place, as none of them had lived philosophically. Finally, what Porphyry explicitly tells us about the prophet and the location of the choice (see below) provides a serious blow to Deuse's reconstruction. We might conclude, then, that for Porphyry as for Proclus, the myth of Er is *not* primarily about the souls that are descending from the intelligible region (even if both were surely interested in providing an account of this, too [cf., e.g., Proclus *in Remp.* 2.185,23-186,1 and Porphyry *Sentence* 29]).

The location of the choice. In the myth of Er, the souls are said to make a twelve-day journey from the meadow to a place from which they could see a column of light as well as the spindle with its eight whorls representing the spheres of the heaven (*Rep.* 616B-7C), and it is here that they receive their lots from the 'prophet' (617D3, 619B3 and C5) and make their choice (619B7-8). Two details here deserve emphasis. First, the souls are said to take in the details of the entire spindle, i.e. of all the celestial spheres. Secondly, Proclus explicitly follows Porphyry in interpreting the enigmatic light here to refer to the vehicle of the World-Soul (*in Remp.* 2.196,22ff. and 185F). Both of these details might lead one to expect Proclus (and Porphyry) to say that the souls are located at the edge of the universe when they make their choice, but Proclus clearly does not think this. Rather, he argues that the physical nature and location of the aether is sufficient to explain how the souls could view both the heavens and the vehicle of the World-Soul (see, e.g., *in Remp.* 2.185,23-188,5 and 198,29-199,21). Likewise, Proclus thinks the aetherial region is an adequate viewing point for reading the lives written in the stars (see below on the lots). The conclusion we should draw from this is that none of these details of what the souls are said to see –the light that is the vehicle of the World-Soul, the spindle in its entirety, the lives written in the stars – would necessarily have forced Porphyry to conclude that the souls are choosing somewhere at the outer edge of the heavens. Thus, pinpointing the location of the choice requires additional textual evidence. Do we have any?

According to Deuse's reconstruction (see especially 151 with n. 177, and 155-6) the first choice takes place either at the outermost edge of the sphere of fixed stars (the sphere of fixed stars has depth, and not all of its stars are the same distance from the centre – see, e.g., Proclus *in Remp.* 2.218,113-16) or altogether beyond the sphere of fixed stars in the so-called 'starless' sphere (there is no mention of this sphere in what we have of Porphyry's works, but it had been postulated by others, including notably Porphyry's student Theodorus of Asine [*apud* Proclus *in Tim.* 3.187,16-21]). His support for this view seems to be as follows. First, this is exactly the location one would expect if the souls in question are descending from the intelligible region (which as we saw above is a questionable assumption). Second, he acknowledges that the fragments make clear that this choice must be made somewhere *within* the universe. This is because the universe's revolutions are said to determine the order in which the souls choose (271F,5-15), which wouldn't make sense if the souls are outside the universe, and because the power of self-determination of disembodied souls making this first choice is said to be 'stained' (271F,16-20), which couldn't possibly be the case if the souls were still in the intelligible world (see Deuse pp. 155-6). Finally, after making their choice the souls are said to move to the corresponding *hôroskopos* (271F,72-5 and 90-5), which would seem to be somewhere on the sphere of fixed stars, since the zodiacal signs are constituted of fixed stars. So Deuse concludes that the souls must have originally been beyond this sphere (or at its other edge) prior to this choice. This downward motion is then continued during the soul's so-called 'first-life' in the planetary spheres (271F,68-70), until the soul finally descends into the sublunary region (cf. 271F,67-8).

The points of tension in Deuse's reconstruction should be apparent. Because he is holding on to the idea that the souls are coming from outside the universe, he has a very difficult time accommodating the textual facts that the souls are both already in the universe and that they are moving towards the sphere of fixed stars. Thus he is forced to posit the starless sphere or else rely on the depth of the fixed sphere itself. Moreover, he runs into problems when trying to account for Porphyry's remarks on when and where the second choice is made. There are indications in the text that the second life is chosen while in the planetary spheres *en route* to the *hôroskopos* in the sphere of fixed stars (271F,50-1,72-5,90-2 and see below), which is simply incompatible with Deuse's reconstruction. Thus, he must resort to collapsing several distinct stages in the soul's journey together.[9] But the greatest obstacle facing Deuse's reconstruction is one of omission. Nowhere does he take into account (or for that matter even acknowledge) 186F (Proclus *in Remp.* 2.255,4-9; 256,9-16),[10] in which Proclus reports that Porphyry identified the 'prophet' with the lunar intellect. Presumably this omission is due to Deuse's considering the fragments on the *Rep.* commentary and the *On What is in Our Power* to belong to different treatises, coupled with the view that Porphyry's exegetical suggestions are not always consistent from treatise to treatise. But as I argued above, there do not seem to be any good reasons for holding the two works to be distinct, and even so, if Porphyry had changed his mind on this matter, Proclus surely would have mentioned it as part of his criticism of Porphyry's view. Now recall that it is the prophet that is responsible for determining the order in which the souls choose (*Rep.* 617D3-4; 271F,12-14), and that Proclus identified this prophet with an aetherial intellect since he maintained that the choices were made in the aetherial region. Porphyry, it seems, has decided that the location in which the souls choose is the lunar sphere, and after making its choice, a soul *ascends* to the *hôroskopos* in the sphere of fixed stars before descending back down through the planetary spheres to the sublunary region and animating a body. This interpretation best accounts for the textual evidence that we have seen so far and is further corroborated by the following.

(i) In the myth of Er the souls, after choosing, are said to be led first to Clotho and then to Atropos, and subsequently under the throne of necessity and to Lethe and the river Ameles (*Rep.* 620D6-621B1). But Clotho, responsible as she is for the moving the outer circumference of the spindle with her right hand (*Rep.* 617C6-7), is easily associated with the sphere of fixed stars, and Atropos likewise with the planetary spheres on account of her moving the inner whorls with her left hand (*Rep.* 617C7-8), as Proclus verifies in both cases (e.g. *in Remp.* 2.267,6-8).[11] Indeed, Proclus is hard put to explain how these remarks about moving to Clotho and Atropos are compatible with his interpretation that the souls never rise above the aetherial region beneath the moon (*in Remp.* 2.341,13-345,26). (ii) Proclus himself appears to associate Porphyry's identification of the prophet with the lunar intellect with the view that the souls are travelling upwards into the celestial region rather than downward into birth:

> [The prophet is] positioned not in the moon, as one person says, but in the aether. For we believe that this place is the starting point of nativity. And the prophet is the guardian of the mortal cycle, and not of the upward journeys into heaven but of the downward journeys into nativity (Proclus *in Remp.* 2.256,12-16).

The last lines (14-16), which Smith does not include in his 186F but which I have added to the translation of 186F below, strongly suggest that Proclus was criticizing Porphyry for thinking that the souls ascend from the lunar sphere into the heavens after choosing a life. (iii) Finally, if the souls really do make their first choices in the lunar sphere, Porphyry's theory of the vehicle of the soul would seem to demand that the souls first move up to the fixed sphere before descending back down into the sublunary region. This is because different kinds of lives would seem to require different kinds of vehicles, and vehicles are constructed on the way down from the sphere of fixed stars (*Sentence* 29). Thus, a soul that has chosen a radically different life would need first to shed the layers of its current vehicle by ascending to the fixed sphere, and then acquire a new one by descending again along a new trajectory.

The lots and choices. Whereas Proclus innovates by introducing two lots, Porphyry brings in two lives and choices that each soul makes prior to each reincarnation. There are many interesting parallels and differences between Proclus's two lots and Porphyry's two lives, but here I shall limit myself here to completing the sketch of the soul's journey through the cosmos in particular relation to its choices and its lot. As shown above, the lots are distributed and the first choices made in the lunar sphere. Porphyry says that the lots are somehow determined by the order in which the souls are carried around the lunar sphere (271F,9-11), though further details are left open, e.g. whether this order is based on merit as in Proclus. Yet Porphyry's report on the Egyptian belief that the first degrees of a sign are generally more beneficent than that sign's final degrees bears some similarities to Proclus' remarks about the sense in which later souls have a more limited choice (*in Remp.* 2.273,6-274,21), which might in turn suggest that the souls' merits are indeed being taking into account. The lives that the soul sees written on the 'celestial earth,' i.e. on the sphere of fixed stars (271F,53 and note *ad loc.*) are simply generic lives of various species and, in the case of human beings, sexes. It is not clear how each of these lives is meant to correspond to the zodiacal circle. What is clear is that since the second life seems to be associated with a specific *degree* of the zodiacal circle (271F,50-7, 75-8, 92-5), the first life would seem to have to correspond to some arc of the zodiacal circle that contains that degree.[12] Porphyry's view appears to be that after choosing a first life (arc on the sphere of fixed stars), the soul proceeds to this arc, but the manner in which the soul proceeds there depends on its own internal state and inclinations These different routes ultimately determine to which degree the soul goes and thus which second life the soul 'chooses' (271F,50-1 and 72-9). Thus, the second life is not so much chosen as provided: it is a matter of justice and, to the extent that the workings of justice are often obscure to the human mind, of chance (271F,78-9; cf. 271F,93-4).[13] This is why Porphyry tells us that 'the choice of lives for the souls outside is principally of the first lives' (268F,82-3), and I believe this is also the sense in which the souls' capacity for self-determination is 'stained by their previous life' (271F,16-17). The most important distinction between these two lives concerns the degree of necessity involved in each. One's species and (in antiquity at least) sex is a hard biological fact that, once chosen, cannot be changed no matter how one pursues one's life (268F,90-5). By contrast, the second life is still in our control to some extent (271F,114-15; cf. 269F,5-6; 271F,29-31). It concerns some non-essential addition (*ti sumbebêkos* 268F,55) to one's species and sex, which Porphyry appears to associate both with one's genealogical starting point as well as natural gifts and talents (268F,64-5) and with one's ultimate career choice (268F,67ff.). The idea seems to be that by 'choosing' its second life, a soul is born to a certain family with certain talents and difficulties, all of which predispose the soul towards a certain career trajectory. Nevertheless, even though one's starting point in life is out of one's hands, one still has the power to resist this trajectory and choose some other career, though this might be very difficult, or else one might pursue the career in different ways (268F,77-81). This is the sense in which the second life is both already determined and still in our power to choose (269F,5-7; 271F,29-31).

With this we might conclude our brief examination of Porphyry's vision of the soul's movement through the cosmos by noting that these results, if correct, effectively refute Deuse's ingenious solution to the difficulties surrounding Porphyry's views on the transmigration into animals. For Deuse (158ff.) wants to reconcile the conflicting evidence by making *On What is in Our Power* a key text on the issue and restricting the first choice (and thus the possibility of transmigration) to souls descending from the intelligible region, and to interpret those passages in which Porphyry seems to deny such transmigration as relating exclusively to souls that did not achieve ascension into the intelligible region. But as we have seen, the textual evidence speaks for such a reconstruction of *On What is in Our Power*.[14]

Notes

1. The only previous translations of this text into a modern language that I am aware of are: (i) a French translation by Festugière, as an appendix to the third volume of Festugière (1970), and (ii) that contained in a modern Greek translation of Stobaeus by Theodoros Mauropoulos (Kaktos 1995). The latter was kindly brought to my attention by Erik Henriksson at a very late stage in the project, and I have not been able to consult it.

2. See Smith *Porphyrius Fragmenta* xiii and his note *ad* P.18 (p. 206), who notes that there is only one fragmentum incertum that would seem to fall outside the scope of the myth of Er on Agamemnon's dream (Homer *Il.* 2.1-34 and Plato *Rep.* 2.383A7-8), but surely many treatises would have offered Porphyry the opportunity to discuss divinely sent dreams.

3. Macrobius only refers to Porphyry by name twice, and in both cases he is clearly referring to other treatises: at 1.3.17 to either *De Homeri philosophia* or *Quaestiones Homericae* (see Stahl *ad loc.*) and at 2.3.15 to his commentary on the *Timaeus*.

4. See my 'The Myth of Er and the Problem of Constitutive Luck', forthcoming in A. Sheppard (ed.), *The Reception of Plato's Republic* (BICS).

5. Here I would like again to express my gratitude to Wolfgang Hübner and Dorian Gieseler Greenbaum for reading and commenting on the astrological passages in earlier version of this translation and introduction.

6. See Deuse (1983) 152n.78. Regarding this relation between the sensible and intelligible universes, see Wilberding (2005).

7. And compare 473F? Smith in which Porphyry might be giving a very different interpretation of the *stomia*. For a thorough examination of these celestial gates, see Hübner (2006).

8. Macrobius *Commentary on the Dream of Scipio* 1.12.3, and see Smith (1993) 206 and Pépin (1974).

9. 'Während bei der ersten Wahl die Bewegung zum Horoskopos eine Folge der Wahl ist, fallen bei der zweiten Wahl die Bewegung zur Gestirnkonstellation, der Wahlakt und die Bewegung durch die Sphären der Gestirne, die die Konstellation bilden, in eins zusammen' (Deuse (1983) 157).

10. Smith chooses to limit the scope of the fragment to *in Remp.* 2.255,4-9; 2.256,9-14, but I believe 2.256,14-16 are very relevant, as I show below.

11. Porphyry might well have interpreted the remaining journey in the manner that Proclus does: To go under the throne of necessity is to enter into the sublunary world which is most subject to necessity, and the water of Ameles signals the humidity that affects one's vehicle (and thus one's cognitive functions). See Proclus *in Remp.* 2.346,1-350,22, and cf. *Sent.* 29.32-43 for Porphyry's agreement on the negative effect of humidity on the soul's vehicle.

12. And see 271F,90-5: The first choice involves going to *ton hôroskopon* while the second choice involves going to <u>*tonde*</u> *hôroskopon*, which might be expressing roughly the same thought. There are obvious difficulties with this kind of correspondence. Given the variety of species of animals, how many distinct arcs corresponding to candidate first-life choices must there be? Also, if human beings are represented just by a single arc on the zodiacal circle, then it would seem to follow that human beings could only be born during a short interval of the day, when that arc was at the eastern horizon, which is implausible (but see 271F,57f. where Porphyry seems to be addressing a similar problem).

13. In fact, as I argue elsewhere (see note 4), this only signals the *beginning* of one's choice of second life, since the choice is only completed by the decisions one makes during one's life on earth.

14. Further problems with Deuse's solution have been raised by Carlier (1998) and by Baltes in his (1985/1999) and in (2002) 264-71 (esp. 268-9).

Porphyry's Commentary on Plato's *Republic* (181-187F Smith)[1]

181T[2]

Many have applied themselves to thinking about this [*viz.* the Myth of Er], including the leading Platonists – Numenius, Albinus, Gaius, Maximus of Nicaea, Harpocration, Euclid, and above all Porphyry, whom I declare to have been a perfect interpreter in particular of all the hidden material in the myth.

182F[3]

The Epicurean Colotes brings a charge against Plato of (i) doing away with scientific truth and wasting his time telling fictitious (*peri to pseudos*) stories like a poet, without demonstrating that he has scientific knowledge [of these things]. And [he charges] that (ii) Plato contradicts himself by on the one hand reproaching the poets in the preceding books for forming frightful [ideas] about the things in Hades and creating a fear of death in their listeners, and on the other himself changing his philosophical tune at the end into a tragic story-telling (*muthologia*) of the things in Hades. For the bellowing mouth, the fiery and wild executioners of the tyrant, Tartarus and all this sort of thing – what form of excess, asks [Colotes], does this leave to the authors of tragedies? And thirdly, that (iii) these sorts of myths necessarily have no real purpose: They are not suited to the masses because they cannot understand them, and they are unnecessary for the wise, as they have no need to become better from such objects of fear. And since [these stories] cannot of themselves tell [us] for whom they are written, they show us that the effort directed at telling them is pointless.

But we should be able to meet these arguments, not only by saying what the preeminent philosopher Porphyry contended, e.g. that a consequence of the proofs of the immortality of souls is that the philosopher also talks about their time in Hades, about the different places there, and about their fortunes in these places, these fortunes being of all kinds in terms of the species of life-form, and these are things that it is absolutely necessary for those to accept who have admitted that souls are immortal, just as those who posit that there is providence consequently must grant the existence of prophecy.

And that Plato did not oppose every form of story-telling[4] but the form that proceeds through shameful and unlawful images, such as Homer and Hesiod composed. Nor does [Plato] frighten the souls and render them fearful of death by setting up these objects of fear in

Hades without provisions (*gumna*). Rather, by presenting these [events] to the unjust, [Plato] makes his listeners hesitant to commit injustice, and he all but draws the conclusion: if being unjust is choice-worthy for you, then the most horrible places of punishment are choice-worthy for you. But you flee these with all your might; therefore, you must also flee injustice.

And that this fiction is in a way natural, because even Nature likes to hide, according to Heraclitus. And just as the *daimons* that guard nature through some such fictions as these reveal their gift to us in the form of dreams or waking visions. And they use ambiguous language, and signify different things through different fictions, revealing images endowed with form (*memorphômena*) as likenesses of things having no form at all (*amorphôtôn*), and still other things through analogous figures. Sacred ceremonies and acts of initiation are full of these things, which actually draw their efficacy from this secrecy and concealment among the initiated.[5]

183F[6]

When Colotes actually illustrates bad taste in terms of the text by objecting to the beginning of the tale for being stylistically immature on account of the parallel use of the words *Alkinou* and *alkimou*,[7] Porphyry was right to attack him for having no taste for rhetorical and sophistical grace and for the Attic sense of style (*mousa*), but at the same time when Porphyry reproaches him because his own teacher had called him 'Colotarion', he did not properly understand Epicurus' own use of an affectionate diminutive of 'Colotes'.[8]

184F[9]

Dismissing, then, the empty conjectures of the interpreters – which Porphyry too rejected – who introduce an immense amount of nonsense on account of the 'twelve',[10] with some referring the number to the zodiac, and others to the months, and others to the twelve gods, believing that in this way [the number 12] is relevant (*oikeion*) to the souls that are going to their nativity – if we must refer this [number 12] to the life of the souls, then we will pursue this theory in a different manner, crediting the kinship of this number to the entire cycle of the souls.

185F[11]

And that this light[12] is something other than the celestial things, has been shown in what has just been said – but now we must say what the nature of this light [really] is. I know that the preeminent philosopher Porphyry suspected as much as we are writing now, having established that it is the first vehicle of the cosmic soul and analogous to the luminous vehicle of our soul[13] and he took a stand for the view that this light is a body that is prior to the celestial body that is fastened to the World-Soul, and surely simpler than the celestial one in corporeal terms. And perhaps the myth in the *Phaedo*[14] also refers enigmatically to this light, when [Plato] says that those who were able to fly up to the upper reaches of the air and having the power to remain there would see that the things there are clearer – 'the true heaven', he says, and 'the true earth'.[15] And in addition to these he adds 'the true light'[16] and in that text he says that there are three things, as you see, heaven, earth and light, and just like here [in the myth of Er] he said that light is a bond between heaven and earth.[17] So then the things that [Plato] postulated hypothetically in the *Phaedo*, here in the *Republic* he has presented as being the case. [...][18] and he makes the souls see heaven, earth and light, which they had not seen before, [...].[19] Be that as it may, as I said, it appears that Porphyry, with these things in mind, determines that this light is the vehicle of the World-Soul and is primarily dependent upon this soul. For it is necessary that prior to the composite body there be a simple body that participates in the incorporeal life in the [soul].

185aF[20]

As for the other proofs (*pisteis*) [*viz.* regarding the nature of place], there is one that is based on the light that is described in the *Republic* as resembling a rainbow, indicating that all of this is meant as a kind of riddle. It might well be in fact that the luminous (*augoeides*) vehicle of the World-Soul is being indicated here, as is the interpretation of Porphyry, but it might well be something else, too.

186F[21]

And for these reasons we do not say that the prophet is the lunar intellect, as Porphyry assumed. For even the lunar intellect is divine and has been allotted a demiurgic rank – it is hardly a servant of the Fates – and it steers the whole realm of generation. Nor do we say that it is the demiurge of the universe, as Theodorus of Asine says [...] It will be clear to you why this prophet in the universe is representative of the divine intellections (*noêseôn*) in Lachesis; it is an intellect, but a particular intellect – an intellect of the Fates (*Moiraios nous*) – positioned not in the moon, as one person says, but in the aether. For we believe that this place is the starting point of generation. And the prophet is the guardian of the mortal cycle, and not of the upward journeys into heaven but of the downward journeys into generation [i.e. nativity].[22]

187F[23]

Let what Porphyry tells us be made known, namely that Plato, having learned about the ascensional times from the Egyptians, indicates that the soul of Ajax has the twentieth place in terms of the risings of the times that determine the lives, and that it was then[24] by directing his attention to the universe that the messenger of these accounts [*viz.* Er] counted the order, I mean [the order] of the souls that are choosing first, second, twentieth or whatever other position. For we too have encountered the Exotic Spheres[25] of the Egyptians and Chaldaeans that determine the differences in lives according to the degrees of the zodiac, making the one degree, maybe, kingly, and the next one – and this is paradoxical to hear – a kind of mercantile degree or one that is worse than even this life (*zôê*), and another degree [they make] that of a priest, and the one after that is of a slave and – what is even worse than this – a man who is without shame regarding his male nature.[26] Thus, it is not surprising that souls that draw lots together (*sungklêrous*) have the first, middle and last [position] in terms of the ascending degrees, since the differences in the lives, too, are in accordance with these, as the spheres mentioned above indicate. One of these spheres, we discover, has †*Kuratos*†[27] inscribed upon it, and another refers to another author (*patera*) of treatises, but all of them, as I said, determine one's life by the peculiarities of the degrees, since the quality in terms of these lives is forecast by the configurations[28] of the planets in relation to the [ascending] degree.

Notes

1. See the introduction on the possible identity of this treatise and the *On What is in Our Power*.

2. Proclus *in Remp.* 2.96,10-15.

3. Proclus *in Remp.* 2.105,23-107,14. This same material is found in Latin in Macrobius' *On the Dream of Scipio* in a slightly different order in 1.1.5-6 and 1.2.3-6 and 11. Smith prints both sources in parallel columns. For a translation of and commentary on the Macrobius passages, see Stahl (1952).

4. The term here is *muthologia*, but I believe the broader 'story-telling' is more appropriate to the points Porphyry and Proclus are making here.

5. On the use of images to reveal divine truths in the Platonic corpus, see Proclus *Theol. Plat.* 1.17,9ff. Saffrey-Westerink (along with their introduction in (1968-1997) vol. 1, clxxi-clxxii).

6. Proclus *in Remp.* 2.111,6-13.

7. A play on words made at Plato *Rep.* 614B2-3. *Alkinou* is the genitive of the proper name Alcinous, and *alkimou* is the masculine genitive singular of *alkimos,* meaning 'brave.'

8. As I now understand this fragment, Proclus, while mostly agreeing with Porphyry that Colotes was wrong to criticize Plato's style, is also correcting Porphyry for misconstruing Epicurus' use of 'Colotarion'. Porphyry presumably understood *Côlôtês* and *côlôtarion* as play on words analogous to that of *Alkinou* and *alkimou* in the *Republic* – poor Colotes' name also had the meaning 'gecko', of which *côlôtarion* would be the diminutive 'little gecko' – but as Proclus points out Epicurus probably just used it as the diminutive of Colotes' name without intending the connotation.

9. Proclus *in Remp.* 2.120,15-24.

10. See Plato *Rep.* 614B6 where Er is described as being 'dead' for twelve days, and 616B2-6 where the souls are said to set out on the eighth day after their return from the places of punishment and reward and to arrive four days later at the pillar of light.

11. Proclus *in Remp.* 2.196,22-197,16.

12. The column of light described in *Rep.* 616B5-6 as 'most resembling a rainbow, only brighter and more pure'.

13. Smith refers here to Porphyry *de Abst.* 2.36.6-37.2; Syrianus *in Meta.* 84,31-85,28; Macrobius *Commentary on the Dream of Scipio* 2.2.14; Proclus *in Remp.* 2.94,8-11.

14. Plato *Phaedo* 109E2ff.

15. *Phaedo* 109E7-110A1.

16. *Phaedo* 109E7.

17. *Rep.* 616C2.

18. Smith marks a lacuna of 21 characters (Festugière: 23 characters), then *kai aga*[*th*].*n* followed by a lacuna of 7 characters, *on,* and a lacuna of 9 characters.

19. The text has *tautais no...n,* and Smith notes that there might be one letter preceding the *no.* This is not enough information to allow for a translation.

20. Simplicius *in Phys.* 615,32-5.

21. Proclus *in Remp.* 2.255,4-9 and 256,9-16.

22. This last line (Proclus *in Remp.* 2.256,14-15) is not included by Smith in 186F.

23. Proclus *in Remp.* 2.318,4-27.

24. Reading *tote* for *touto* with Usener and Festugière in 187F,5.

25. I would like to thank Dorian Gieseler Greenbaum for pointing me to two discussions of this passage: first, in Boll (1903) 364ff., who describes the *Sphairai Barbarikai* as a volume containing not only the stars and constellations of the Egyptians and Chaldaeans, but also their effects on the lot in life that one who is born under them receives;' and secondly, in Housman (1930) xl-xlii.

26. Compare this with 268F,90-1. Festugière, following Kroll, directs us to Firmicus Maternus 8, 5, where he claims that *cinaedi* are said to correspond to the tenth degree, but as Wolfgang Hübner has kindly pointed out to me, there are no 'cinaedi' in 8, 5. The long passage on the *cinaedi* is 7, 4-16, but here there is no mention of the tenth degree.

27. Smith prints *Kuratos* – the manuscripts apparently originally had *kratos* – but encloses the name in daggers. Boll (1903) 369 follows a proposal by Schoell of *Kuranos,* which was doubted by Kroll (1924) on the grounds that the Cyranides were not concerned with astrology. Festugière (*ad loc.*), however, is able to report a case of 'Kyranides astrologiques', though he remains cautious and prints only 'intitulée "Sphère de K..."' in his translation.

28. As Dorian Gieseler Greenbaum has pointed out to me, *suskhêmatismos* refers to the entire astrological/astronomical configuration of the chart, the whole layout of the planets within signs of the zodiac, as opposed to aspects, which are specific geometric connections between two planets.

Porphyry
On What is in Our Power (268F-271F Smith)[1]

268F

In the discussions we had with a little while ago, Chrysaorius,[2] we brought up the issue of the choice of lives, and that Plato, by assigning it to the souls that are outside [of their bodies],[3] seems at risk of doing away with what is in our power (*to eph' hêmin*) and in general with our so-called power of self-determination (*to autexousion*),[4] if it is really the case that they come to this choice from the lives that they lived before during the previous cycle[5] with their characters already formed by the things that they loved or hated, and by what caused them pleasure or pain. It seemed right to us to leave these problems aside, since we believe that it is possible to provide some convincing points in their defence.[6] But as for the things that [Plato] says the souls, through entering their bodies, suffer after the choices are made and are destined (*mellein*) to appear as particular human beings, these things appeared difficult to us. For at one point Plato says:

> 'Let the first choose a life (*bios*) to which he will be joined by necessity'[7]

and at another point says that the *daimon* that we had been allotted is a kind of inescapable warden for us.[8] For [Plato says] Lachesis, who is a daughter of Necessity, sends with the one who received a lot and chosen some life, a guard of this life and a complete [*ek pantos*] fulfiller of what had been chosen. And this [*daimon*], having received the human being, leads him to Clotho, under her hand and the rotation of the spindle, and after this to Atropos, who makes the destiny that has been spun irreversible, and without turning around he directs him from there directly under the throne of Necessity, and once he has made him pass through that, he leads him, and the others at the same time, to the plain of Lethe beside the river Ameles, from which one must drink a measure [of water], and those who drink utterly forget everything. Thus, since these things have been assigned as well as necessitated and ratified by the Fates and Lethe and Necessity, with a *daimon* accompanying them and ensuring their fate, *what* could we be in control (*kurioi*) of? And how is it still the case that

'Virtue knows no master. Each man, by honouring or dishonouring virtue, shall have more or less of it'?[9]

So, when I took the [works] of Plato into my hands and saw that we were seriously mistaken as to the philosopher's meaning, I was surprised at us[10] for having been at a loss. For whoever is working on the assumption that he is discussing 'lives' only in one sense – the one in which those who give systematic treatments of the matters of lives [use it] and make mention of a kind of farmer's life, and again a life of a politician, and another life of a soldier – and supposes Plato to be using the term 'life' in the same way, this person is justifiably perplexed over how a man who has chosen one of these lives is forced to live out this life alone in its entirety, since many different types of life are observed among human beings and the choice of them is in our power in most cases – as is the amendment of one's [previous] choice of them. The Stoics, for example, assume that 'life' is meant in just one sense – that of a *rational* life-form – which they further expound as a discursive passage (*diexodos*) consisting of activities, relations, doings and sufferings, but Plato calls the life-forms (*zôas*) of non-rational animals 'lives' (*bious*) too. For the life-form of a swan is also a life according to Plato, and again the life-form of a lion, by which one is a lion, is another life, and that of a nightingale another. The human life-form is a life, too, and the one [form] of this life is feminine, while the other [form] is masculine;[11] for Plato says that the paradigms of lives are of all sorts. This sort of thing, then, is one meaning of 'life' in Plato, but there is another, which Plato presents as a sort of non-essential addition to these and a kind of second character on top of the principally-chosen life just mentioned. For the principally-chosen life of a dog is the life in accordance with the life-form of dog, but upon this kind of life there is a non-essential addition of the [character] of[12] hunting dogs or tracking dogs or table dogs or watchdogs, and this would be a second life. Now with non-rational [animals], since they are deprived of self-determination, this character is formed as an addition either naturally or as a result of training by its owner, whereas in the case of human beings, coming from a good family or being endowed with bodily beauty is provided by nature or by chance, except that clearly none of these things was in our power. By contrast, the acquisition of skills and professions and knowledge,[13] as well as the acquisition of political lives and the pursuit of positions of power and all such things – these things happen to depend on what is in our power, even if some [of these things] are difficult to achieve because they depend on our receiving a certain contribution from the outside world, for which reason they are [in some cases] difficult to attain and [in other cases] not easy to turn one's back on, e.g., positions of power and leadership and tyrannies. All of these sorts of things happen to depend on deliberate choice, although achieving them is not completely in our power, and yet once (*aph' hou*) we achieve some [life], it is necessary to follow its prescription (*taxis*). Nevertheless, for nearly everyone, if they maintain moderation[14] and have not been reduced to vice, it has been granted to use the lives of which they partake in either a good or evil manner.[15] For even some human beings who were induced to choose tyrannies, turned out to be kind and gentle while others used their kingships wickedly.

Given that this is how things are, the choice of lives for the souls outside is principally relative to the first lives, and then, once that has been chosen, Necessity confirms its inevitability and the *daimon* attends to it, each *daimon* serving, as it were, as a guard and watcher over each life, binding each soul to abide in its chosen [life] and not to abandon[16] it, and the Fates spin out this kind of life for each one, and what happens throughout one's life bears witness to this.[17] For a man, even if he should become excessively soft on account of licentious behaviour and by deliberate choice changes all of his external features into those customary of the female sex, could never *become* a woman, nor could a woman ever become a man by trying to perform manly activities.[18] For souls live this life by necessity, though they live out this life in forgetfulness (*dia lêthês*) and complete oblivion of the fact that the choice of this life was once in their power. And in the case of human beings, one's power to choose (*to ethelousion*) goes on in the freest manner with respect to matters of the soul, when it is by itself and not bound in the body, whereas with respect to the matters of the living thing, i.e., of the composite being, one's power to choose has grown slack with respect to self-determined actions.[19] For each of us lives out his or her human life with this, namely the possession of an impulse that is appropriate to the self-determined actions *of the composite being*.

269F

For each living thing manages itself according to the life-form[20] that is appropriate to it here, but the life of the human being has in his or her own power the capacity for self-determination that corresponds to a human being. Therefore, even if [a human being] is led by natural impulses[21] to one of the lives that fall within his choice here,[22] not falling into this life was in his power, even if he does fall in completely.[23] For in the cases where we consent to our persuasive power of representation on account of its persuasiveness, still not consenting was also in our power, provided that (*ho ti mê*) the power of representation does not drag us and manipulate us like puppets to itself.

270F[24]

For the whole thrust of Plato's theory appears to be something like this: Souls, prior to falling into bodies and different lives, have the power of self-determination for choosing this or that [first] life, which they are to live out with a certain life-form and a body appropriate to that life-form. (For it is in their power to choose the [first] life of a lion or of a man). That power of self-determination, however, is hindered as soon as they fall into one of these kinds of [first] lives. Once they have gone down into bodies and become souls *of living things* instead of *unconfined* souls, they bear [only] the power of self-determination that is appropriate to the constitution of that living thing, and in some things [this power of self-determination] will be highly intelligent (*polunoun*) and contain much flexibility (*polukinêton*), e.g., in human beings, while in others [it will] be relatively inflexible and one-dimensional (*monotropon*), as in the case of nearly all other living things.[25] The power of self-determination depends on the constitution [of the living thing's body]; although this power has its own source of motion, this motion is delivered (*pheromenon*) against the background of (*kata*) impulses that originate from the [body's] constitution.[26]

271F

In every case it is proper to flee excesses and to pursue the mean,[27] unless what is in our power (*to eph' hêmin*) is already bound and enslaved by the passions connected to vice, and made subject to these.

It is his view (*areskei*) that the souls are neither allotted nor receive their lives in a disorderly manner, but rather in an orderly manner, i.e., as the revolution directs them. For the first soul is the first one to have arrived at a stop in the revolution, whereas the second soul to stop will be allotted the second [position].[28] For as the souls are made by the universe to go around and stop, they are led around by its motion as if in order, with the lots signifying 'first' and 'second.' This is why what are received from the prophet[29] are the lots (*klêroi*) of Lachesis, who has after all received her name from *lankhanein* [meaning 'to obtain by lot'].[30] And many have said that Lachesis is the revolution of the universe.[31]

It is also his view that the capacity for self-determination that [disembodied] souls have in their power is stained by their previous life that seeped into them (*enginomenês*) here, while the capacity for self-determination that [embodied] human beings have in their power [is stained] by the inclination of the soul which is still outside to one of the lives here, and Plato calls this inclination a 'choice' (*hairesis*). And it is his view that this is the sort of thing that fate (*heimarmenê*) is: it is like the prescriptions of laws, it itself being a law, and, as Empedocles says,

> 'an ancient decree of the gods,
> eternal, sealed by broad oaths.'[32]

For laws also prescribe without forcing one [to do anything] – for example, if you are a thief, you will suffer this; and if you are excellent, you will attain these things – and so do the ordinances according to fate – if you choose the [first] life of a man, you will live in this way. But they certainly do not in general (*ek pantos*)[33] *force* [you to choose] <what> you choose.[34] For instance, if ever, having come to be among men, you should choose the life of a soldier, then it is necessary for you to do and suffer such and such, but it was surely not also necessary that you choose the life of a soldier and [thus] do and suffer these things.[35] For this reason there are established paradigms of both the first and second lives. It is in the souls' power to choose the *first* lives, and to live, say, the life of a human being, as well as to choose one of the *second* lives, but once one has chosen [either life][36] and is living according to this life, it is necessary for him to do and to suffer accordingly.[37]

If, however, it is possible for one who is trying to divine Plato's meaning to make a lucky guess as to where Plato took these [ideas] from and discussed them in the belief that they were in complete agreement with the facts, I suppose I would say that [he has them] from Egyptian wise men,[38] who interpreted[39] the lives from the ascendants and from the arrangement in the nativities of the planets with respect to the stars in the zodiac,[40] as consequent to the rising of the ascendant.[41] [However, I would say that] Plato does not agree with the Egyptians that a certain sort of arrangement of the overall configuration (*skhêmatismoi*)[42] *necessitates* that the lives be of a certain sort for the souls that are going into nativity through the degree ascending over the eastern horizon, on account of the aspects at that time being configured in a certain way. Rather, it is in accordance with their internal arrangements (*diatheseis*)[43] that the souls are being delivered to the ascendants and to the lives that the aspects *signify*, as [the souls] behold [the lives] written on the celestial earth[44] as if upon a board.[45] The souls that have chosen are able – on account of their power of self-determination – also not to live in this way, but since they have chosen the things that are intersecting in the revolution at the point of time of the ascendant's rising, then they must necessarily live out [at least] what is written.[46]

Why, then, in the same ascension is, say, a dog generated and a man and a woman and many men,[47] for all of whom neither the first life nor the second life is the same? First, I would say that for all the souls the journey is not so much faster that the rising of the ascendant fails to precede their entrance [into the body], and since the revolution (*phora*) always comes before and since each difference in lot delivers a different paradigm of life, those who *appear* to charge into the world at the same degree of the zodiac (*hôra*),[48] are necessarily never exactly coinciding.[49]

Secondly, the lot *exô* ††, until it has fallen into the sublunary region.[50] The pathway of the first life takes place across the seven spheres, but each [soul] moves through these [spheres] in a different way according to their impulses towards certain second lives, and so this pathway does not make all souls devote themselves to the[51] written [lives] in the same way. Thus, the soul that has chosen the life of a dog immediately goes to *this* ascendant, while the [soul that has chosen the life] of a human being goes, by turning aside from that point, to *this* ascendant.[52] And Justice (*Dikê*) delivers the soul in accordance with the individual formation of its character[53] to a certain degree of the revolution that has a life written [on it] that befits that soul's impulses, but Justice is called Chance (*Tukhê*), because it is a cause that is obscure to human reason.[54]

There are twelve signs of the zodiac, through which, as the Egyptians almost universally believe, the path leads all souls to come to be down here [on earth], and the first degrees of each zodiac sign[55] were traditionally thought to be of large extension,[56] as if they had been allotted to the governor of that sign.[57] By contrast, the final degrees in every [zodiac sign] have been assigned to the so-called maleficent stars.[58] Hence, the positive endowment (*eumoria*) of the first lots (*klêrôn*) is deemed worthy of acceptance, and the positive endowment of the last lots is said to be limited.[59] Plato, then, establishes that configurations of a certain sort *signify* the lives, but they do not *necessitate* them.[60] Rather, the souls that have made their choices to live just as [the configurations] signify must necessarily take hold of the concatenation of written [consequences]. Thus, the cause of their moving to the ascendant belongs to their having chosen the [first] life of a woman or a man or some other living thing. But the cause of the soul's having been carried to this [particular][61] ascendant belongs to[62] its having chosen the second life, which the order of the planets and their arrangement with respect to the ascendant reveals in written form. Unless one says that the observation of the ascendant at conception makes clear that one has chosen the lot of human being or dog, while the observation of the horoscope[63] at the moment of birth displays the choice of second life that comes in addition to the one already chosen.[64] But one might concede that this story-line is drawn from the traditional views of the Egyptians,[65] but it is difficult to show whether parts of [this story-line] were revised by Plato in terms of its conceptual content.

Since you happen to be a lover and admirer of Homer, too, take a look for me and see whether he, too, even before Plato, acknowledges the dual [choice] of lives – the one being a kind of unchangeable life that the soul can either choose or not choose, but once it has chosen, although it is then unable to escape and change this choice,[66] it regulates this life through either virtue or vice. He reminds us of this when he says:

'I declare that fate is not something to be evaded by any man,
Whether he be base or noble, once the first things take place'[67]

This, then, is what Homer might be saying about the first, unalterable life, but what does Homer say about the second life, which is something in our power?

'How mortals do blame [us] gods!
For they say that evils come from us [gods], but in fact they themselves also
By their own follies bear pain beyond their fair share.'[68]

Why, then, [does Homer say] 'they themselves *also*'? For he is making
clear that from the gods, too, some [bad] comes to mortals, even if the 120
mortals[69] are themselves responsible for the majority. Yes, because the paradigms of lives come from the gods, even the gods are not situated beyond all responsibility for our falling into evil.[70] And yet he acquits the gods by making the souls have the power of self-determination and entrusting to them the choice of lives, and those who choose hastily and ignorantly[71]

'By their own follies bear pain beyond their fair share.'[72]

Notes

1. On the possible identity of this work with Porphyry's commentary on the *Republic*, see the introduction.

2. Chrysaorius, a friend of Porphyry's and a senator in Rome, is also the addressee of Porphyry's *Isagoge*. The little we know about him is briefly discussed by M.-O. Goulet-Cazé (1994) 323-4.

3. Are the souls just outside their bodies, or outside the entire universe? This is a critical point of interpretation. Deuse (1983) assumes that the latter and takes the following discussion to be about what happens to souls that are entering the visible universe from the intelligible universe, but there is good reason to be sceptical of Deuse's assumption. Most, if not all, of the souls described in the myth of Er do not appear to have lived the kind of life that would allow them access to the intelligible world after death, since this would seem to require living not just a just life but a philosophical life (cf. *Rep.* 619C7-8, 620A2-3). For this reason Proclus distinctly rules out that these souls are descending from the intelligible region (*in Remp.* 2.161,14-16; 2.185,23ff.), but rather from the aether that is located in the highest part of the sublunar region, just beneath the moon (*in Remp.* 2.131,27-132,5; 2.133,4ff.). Given Porphyry's placement of the prophet in the lunar sphere (186F), it seems likely that the souls in question are also located there.

4. *autexousios* is common in Plotinus (cf. Sleeman and Pollet (1980) *ad loc.*) for whom it is virtually synonymous with *hekousios* (see Atkinson (1983) 8) and with *eph' hêmin* (see Remes 2007). In the *Enneads* it has been translated in various ways, including 'free-will' (Atkinson) and 'independence' (Armstrong (1966-88), Brisson-Pradeau (2003)). I prefer 'self-determination' (as do Beutler-Theiler (1956-71), Remes (2007) 179ff., Sorabji (2000) 334, *et al.*) to avoid some of the modern connotations of 'free-will.' Plotinus notably points to it as a cause of the soul's fall in 5.1.1.5, but on this see D. O'Brien (1977).

5. In *Rep.* 620A2-3 Plato says that 'for the most part' (*ta polla*) the choice of life depends on the character of one's previous life.

6. The contrast here is between the apparent lack of freewill that we have during the choice of lives itself on account of the influence of the past lives over this choice (see above note), and the lack of freewill we have during life on account of the inescapability of the life we have chosen. Porphyry is interested only in the latter here and suggests that the former problem can be dealt with. Presumably what he has in mind is that Plato only says that former lives influence one's choice 'for the most part' (620A3) and that it is possible to choose 'with one's intellect' (619B3). See J. Wilberding, 'The Myth of Er and the Problem of Constitutive Luck'.

7. *Rep.* 617E2-3, slightly modified. The original says: 'Let him who received the first lot choose a life first to which he will be joined by necessity'.

8. What follows is a paraphrase of *Rep.* 620D6-621B1.

9. *Rep.* 617E3-4, slightly modified. Porphyry has *hê aretê adespotos* for Plato's *aretê de adespoton*, but there is no difference in sense.

10. Reading *ethaumasa hêmas* for *ethaumasamen*. Heeren suggests *ethaumasa men*, but the *men* just seems out of place here. Porphyry is clearly criticizing his own past state of puzzlement.

11. Smith refers to *Rep.* 618A.

12. Reading *ho* for *ê* in 268F,59.

13. *tekhnai, epitêdeumata* and *epistêmai* are also in series elsewhere in Stobaeus, e.g. 1.41.6.98-9 and 1.49.44.517-18, and *epitêdeumata* and *epistêmai* are often in series in Plotinus 1.6.4.8 and 5.9.2.6-7 (derived from Plato *Symp.* 210C6).

14. See also 271F,3.

15. Smith includes the following comparable passage from Simplicius' commentary on Epictetus' *Enchiridion*: 'And the souls choose their lives based on their previous dignity (*axian*) and state of living (*zôên*), but they have it in their own power (*exousian*) to use these lives in a good or evil manner. Thus, even some who chose a mercantile life often live in a good manner, and some who appear to be philosophers disgrace themselves. Therefore, the forms of the lives are these, e.g., that of a farmer, or of a merchant, or of a musician, and the souls themselves choose based on their previous state, i.e. (*kai*) the universe determines [this life] for them based on their dignity, but the souls establish by themselves the quality of this life' (46,6-17).

16. Reading *proleipein* (abandon) with Heeren for Smith's *problepein*. Hense and Festugière have *periblepein*.

17. Reading *ta te gignomena dia*... as in MSS for Smith's <*hoion*> *ta* [*te*] *gignomena dia*....

18. This should be read together with what Porphyry says in *de Abst.* 4.20.3 and *Marc.* 33. In the latter passage Porphyry encourages his wife Marcella to flee all that is womanish and to behave as if she were a male. In the former passage Porphyry warns against 'the inner male' becoming feminized by licentious acts. Of course both of these passages are compatible with his point here that certain biological facts relating to sex remain unchanged, even when one alters one's personality.

19. Reading <*pros*> *autexousia erga lelumenon* with the MSS for Smith's (following Heeren) <*pros*> *autexousia erga lelumenou*. I translate the *lelumenon* as agreeing with *to ethelousion* in *l.* 98.

20. In *AG* 12.14 Porphyry uses this term (*diazêsis*) to distinguish between plant and animal life. Here it seems to be equivalent to *zôê*, with both referring to the kind of *first* life, i.e. the species – and, in the case of humans, the sex – of animal.

21. Regarding *prothumiai* see Plotinus 4.3.13.19 in the context of automatic pre-reflective impulse contrasted to *prohairesis*. See also 4.4.20-1 where *prothumia* is distinguished from *epithumia* as preceding ordinary desire and belonging to the qualified body (and cf. 4.8.7.9 and 5.3.12.29-31).

22. That is to say, once one has made the choice of first life and become either a male or a female human being, one is subject to various non-rational impulses towards a certain kind of second life. The kind of second life that one is drawn towards, e.g. the life of a tyrant or that of a farmer, as well as the strength and variety of one's non-rational impulses, will vary from person to person, but every human being has the power not to pursue the kind of life towards which one's baser nature draws one. Porphyry is drawing a strong contrast between pre-rational impulse (*prothumia*) and rational choice, as does Plotinus at 4.3.13.18-20. This passage as well as 270F,17-18 make clear that these *prothumiai* are due to and will thus vary according to the constitution of one's body.

23. Porphyry's use of *empesein* here should be compared to Plato's use of *ekpesein* in *Rep.* 495A5, where what is under discussion is how one with a philosophical nature can 'fall away' from the pursuit of a philosophical life on account of external factors.

24. This fragment is also translated in Baltes (2002) 66 with commentary on 266-70.

25. This should be compared to Porphyry's discussion of animal rationality in *de Abst.* book 3, where he argues that animals are not completely deprived of reason and that it is rather a matter of 'more and less' (3.6.7; 3.7.1; 3.8.7; 3.18.1; 3.23.8; etc.). He also gives examples of animals overcoming the appetites when they recognize a baited trap (3.14.1). See also Plotinus 6.7.9.13-14 and Damascius *de Princ.* 1.32,19-23 Westerink, where the *hormai* of non-human animals are called *monoeideis* and *autophueis*.

26. On the bodily impulses, see 269F,5-7 with the note *ad loc.*

27. cf. *Rep.* 619A5-7. And see 268F,77-8.

28. The sense of the Greek here is difficult: *hê men gar prôtê pausamenê en têi prôtêi periodôi prôtê hêxei, hê de deutera <deutera>* (added by Smith following Wachsmuth) *en têi deuterai klêrôsomenê*. I find the *en têi prôtêi periodôi ... en têi deuterai* problematic, since the fastest celestial revolution would be that of the sphere of fixed stars which completes a revolution every day, but there is surely more than one soul a day going into generation. Other passages (187F, 271F,79ff. Cf. Proclus *in Remp.* 2.265,15-23) express more clearly that the order of lots is determined not by the *number of revolutions* but (in some manner) by the *degrees* of the revolving circle. For this reason I read *hê men gar prôtê pausamenê en têi* [*prôtêi*] *periodôi prôtê hêxei, hê de deutera en têi deuterai* [scil. *taxei*] *klêrôsomenê*, which is no more intrusive than Smith/Wachmuth's addition of *deutera*.

29. Plato mentions the prophet who distributes the lots at *Rep.* 617D3, 619B3 and C5. On the identity of this prophet, see Porphyry 186F.

30. The text as it stands has *hêtis apo tou lankhanein kai klêrousthai keklêtai*, but the *hêtis* must be referring to Lachesis alone, whose name is surely connected to *lankhanein* but certainly not to *klêrousthai*. Thus, I have decided to bracket *kai klêrousthai* (also meaning 'to obtain by lot'), which is clearly aimed at (correctly but unnecessarily) explaining *klêroi* ('lots'). Cf. Proclus *in Remp.* 2.273,9. On the specifically astrological meaning of the term 'lot' (*klêros*), see now Greenbaum (2009) ch. 5.

31. Proclus makes Lachesis the whole heaven, including both the fixed stars and the planetary spheres, since Lachesis is moving the spindle in both directions. If my reconstruction of the path of the soul's astral journey in the introduction is correct, then Porphyry himself might think that it is specifically the revolution of the lunar sphere that is determining the lots.

32. Empedocles DK 31 B115,1-2.

33. For this sense of *ek pantos,* see Strange (1992) *ad* 82,33.

34. Reading *ho helêi* for *helou* in 271F,29, since the imperative *helou* strikes me as rather awkward, though possible. Cf. Baltes' translation of the text as it stands: 'Doch darüber hinaus nötigen sie nicht dazu: "Wähl' doch [dieses Leben]!"'

35. Smith and Baltes refer to the following in this connection: Plato *Tim.* 41E2f.; Apuleius *de Platone et Eius Dogmate* 1.12.205; Ps.-Plutarch *de Fato* 570C and Calcidius *Commentary on Plato's Timaeus* 182,11ff.; 184,21ff.; 187,7ff., as well as to Theiler (1966) 84-5.

36. Once either life is chosen, the consequences are necessary, as Porphyry makes clear in the case of the first life at 271F,28 and in the case of the second life at 271F,29-31.

37. cf. Olympiodorus' *Commentary on Plato's Alcibiades* 45,5ff.: 'It is in our power (*en hêmin*) to choose this kind of life, but once we have chosen, doing what follows from this life is a matter of necessity and not in our power, just as it is in our power to throw a stone, but once it has been thrown, it is not in our power to stop it.' See also Kobusch (1976) 100.

38. 'Egyptians' and 'Chaldaeans' are terms that in late antiquity were commonly applied to astrologers generally (Gundel (1972) 492,5-9).

39. Wolfgang Hübner has suggested to me that Porphyry's use of *sêmeioumenôn* here might be an allusion to the technical term *sêmeion*, which is a point in the sense of a crossing of two lines, as opposed to *kentron*, which refers to a point in the sense of the middle of a circle or a sphere. See Federspiel (1998).

40. cf. Stobaeus 1.5.16: 'Fate is the cause of the arrangement of the stars (*heimarmenê de aitia tês tôn astrôn diatheseôs*)'.

41. Dorian Gieseler Greenbaum has suggested to me that *hôs akolouthous têi tou hôroskopou anatolêi* is included because the way the planets and stars fall in one's astrological chart, and their subsequent relationship, is dependent on where the ascendant falls.

42. On the term *skhêmatismos*, which seems to have the same meaning as *suskhêmatismos*, see the note on 187F,23.

43. The term *diathesis* here recalls the *diatheseis* of *ll.* 44-5, where it clearly had the somewhat different sense of the 'arrangement' of the stars and planets in nativity. It seems unlikely that the term reappears so soon by chance. Porphyry appears to be correlating the 'internal arrangement' of the souls to the external arrangements of the lives laid out from each of the ascendants, and the particular arrangement of the planets and zodiac sign within that particular chart. Thanks to Dorian Gieseler Greenbaum for helping me see this point.

44. In *Peri Agalmatôn* Porphyry reports that the Egyptians referred to the moon (not the entire lunar sphere) as the 'celestial earth': '[The Egyptians] call the power of the celestial earth and the terrestrial earth "Isis" on account of their equality, from which justice [is derived]. They say the celestial earth is the moon, and the terrestrial earth is the fruit-bearing earth in which we reside' (360F,25-9 Smith). This is a practice that Proclus associates with the Pythagoreans and the Orphics, and with which he appears to be in agreement (*in Tim.* 2.282,11-13; 3.142,12-14; 3.172,18-23). At *in Tim* 2.48,15-26, however, he reports that the Pythagoreans and the Orphics refer to the moon as the '*aetherial*' earth, with the '*celestial*' earth referring rather to the sphere of fixed stars, and the sphere of fixed stars seems to be the more likely meaning here (see the Introduction). See also Hübner (1990) 272.

45. As Dorian Gieseler Greenbaum has pointed out to me, *pinax* is the technical term for the astrological board used by astrologers to lay out charts for their clients. See J. Evans (2004).

46. Deuse (1983) 152n.78 and Smith think that the text as it stands is unintelligible. The perceived problem is that Porphyry says that the souls that have chosen both are required to live out that life *and* yet have the power to live differently. To avoid this apparent paradox, Deuse inserts *mê* before *helomenas* in line 54 ('the souls that have <not yet> chosen are able ...'), and Smith more cautiously marks *helomenas* as problematic and suggests *hairoumenas* ('the souls in the process of choosing are able ...') in the critical apparatus, which strikes me as not significantly different from Deuse's suggestion. But I think we can make sense of the text as it stands. The idea would be that by choosing the (first) life, e.g. of a woman, you are required to live out your life in a female body (this is all that is 'written'), but you still have the power to pursue a masculine *manner* of life. This is exactly what Porphyry tells his wife Marcella: 'Regardless of whether you are a man or a woman, do not be concerned with your body, nor should you see yourself as a woman, because I don't even see you as such. Flee everything that is feminine in your soul, as if the body encompassing you was a man's' (*Marc.* 33). As Porphyry emphasizes in 268F,101-4; 269F and 270F,9-18, embodied human beings still have a significant amount of power of self-determination.

47. Wachsmuth suspects, perhaps rightly, that the text is corrupt here. He suggests *athroôs* for *andres*: '... and many all at once ...'. Festugière cautiously suggests *polloi anomoioi* 'many people who are unlike one another'. On the traditional argument that many men of different character or destiny are born at the same time see: Long (1982) and Hübner (1975) 123.

48. The Greek term *hôra* can just mean 'time' or 'hour' (cf. *de Abst.* 4.11.6) but a more specific astrological sense seems appropriate here. The Greek term *hôra* can mean, especially in poetry, the ascendant, and in this sense it is synonymous to *hôroskopos*. See Hübner (2001).

49. This is difficult. I take it that Porphyry's point here is that since time elapses between the soul's going through the horoscope (and thus choosing a life) and its entrance into its human body (i.e. the birth of the human being – see the *AG, passim*), we can never *fully* determine a person's ascendant. Thus the objection about very different people having the same ascendant is misplaced.

50. Smith (following Wachsmuth) signals a lacuna in the text here.

51. The Teubner text has a typographical error here, printing *tous gegrammenois* for *tois gegrammenois*.

52. The text has *kata tên paraklisin tês stigmês*. *paraklisis* is extremely rare, though not a *hapax* (Orphica *Fragmentum astrologicum* 12.160,9), and does not even appear in the LSJ, but if *paraklisis* is right, it has to mean the 'bending' or 'turning aside' of the soul during its downward journey to a different degree. I am uneasy about the *tês stigmês*. It might be a genitive of separation, with the 'point' in question being the degree to which the soul that chose the dog's life went, or perhaps a genitive of measure ('turning aside *by a* [*single*] *point*'), though the definite article makes that unlikely.

53. cf. 268F,8.

54. cf. Proclus *in Remp.* 2.266,31 and 2.294,4-295,27. As Smith points out, this is the Stoic definition of chance. See *SVF* 2.965-7 and 970-1.

55. Reading *hekastou zôidiou* for *tou zôidiakou* as suggested by Stephan Heilen (2010) 82n.92.

56. 'of large extension' translates *amphilapheis*, following Heilen's paraphrase (2010) 58.

57. Reading *autôi* with the MSS for *agathôi* (Smith and Heeren), as suggested by Heilen (2010) 82n.92.

58. See Heilen (2010) 58 and 82n.93 who identifies these stars as Mars and Saturn. Cf. Gundel (1950).

59. cf. Proclus *in Remp.* 2.273,6-274,21, and see the Introduction.

60. This is a commonly held view among Neoplatonists (Plotinus 2.3.1.1-2; 2.3.7.1ff.; 3.1.6.22-3; Proclus *in Remp.* 2.292,13-16), but this is not to say that the stars have no causal efficacy at all (e.g. Proclus *in Remp.* 2.292,26ff.). For a good discussion, see Adamson (2009).

61. The contrast is between *ton hôroskopon* and *tonde hôroskopon*. I suspect that we should understand the latter as referring to some arc of the zodiacal circle, and the latter referring to some degree on that arc.

62. Reading *tou* for Smith's *to* at 271F,93 with Heeren, Boll, Festugière and Deuse (150n.73).

63. Deuse might well be right (154n.86) that this is not the *act* but rather the *object* of observation, i.e. 'Tierkreiszeichen *und* die Gestirnkonstellation, die zu diesem Zeichen gehört, d.h., das Horoskop in unserem Sinne als Stellung aller Gestirne zueinander (zu einem bestimmten Zeitpunkt)'. For a good discussion of conception-based astrology in antiquity, see Frommhold (2004).

64. This appears to show that Porphyry is not completely committed to all of the astrological details of this section. Compare this with 271F,90-5.

65. Colotes had identified Er with Zoroaster and accused Plato of plagiarism, and Porphyry was concerned to refute Colotes' objections to the myth of Er. See Proclus *in Remp.* 2.109,4ff. and Festugière's note *ad* 2.109,16.

66. Smith signals that there is a problem in the text here, but this must be the sense, achieved perhaps, as an anonymous vetter proposed to me, by inserting *hôste mê* after *ekhousa*.

67. Homer *Il.* 6,488-9. Smith refers to *SVF* 2.925 and Theiler (1966) 75.

68. *Od.* 1.32-4.

69. Reading *autoi* for *autous* as Wachsmuth suggests.

70. Wachsmuth marks this as a question in his edition of Stobaeus' *Anthology*, but I follow Smith's punctuation here.

71. *Ex epidromês kai amathias*. Cf. Plato *Rep.* 619D4-5 *ex epidromês* and 619B3 *sun nôi*.

72. *Od.* 1.34.

English-Greek Glossary

to abide in: *emmenein*
to accept: *lambanein*
acceptance: *apodokhê*
to accompany: *hepesthai*
account: *logos*
acquisition: *analêpsis*
to acquit: *apoluesthai*
to act or perform: *dran*
act of initiation: *telestêrion*
action: *ergon*
activity: *energeia*
to add: *prostithenai*
addition, to be a non-essential: *episumbainein*
addition, to be formed as an: *episunistenai*
to address: *prosiesthai*
admirer: *epainetês*
to admit: *paradexesthai*
aether: *aithêr*
air: *aêr*
to be allotted: *lankhanein*
ambiguous: *loxos*
amendment: *metathesis*
analogous: *analogon*
ancient: *palaios*
to appear: *eoikein, phainein*
appropriate: *oikeios*
argument: *logos*
arrangement: *diathesis*
to ascend: *hôroskopein*
ascendant: *hôroskopos*
ascension: *anaphora*
of ascension: *anaphorikos*
to assign: *aponemein, epiklôthein*
to assume: *hupolambanein*
to attack: *epirapizein*
to attend to: *epakolouthein*
author: *patêr*
to avoid: *apotithenai*

to bear: *pherein*
to bear witness: *marturein*
beauty: *kallos*
to become soft: *malakizesthai*
beginning: *eisbolê*
to behold: *blepein*
to believe: *oiesthai*
to bellow: *mukasthai*
to bestow: nemein
better: *ameinôn*
to bind: *dein*
birth: *hê ek gastros ektropê*
to blame: *aitiasthai*
body: *sôma*
bond: *sundesmos*
to bring a charge: *enkalein*
to bring up: *anakinein*
to call: *kalein*
cause: *aitia*
to cause pain: *lupein*
to cause pleasure: *hêdesthai*
celestial: *ouranios*
Chaldaean: *Khaldaios*
to change: *metaballein, metaskêmatizein, metharmozein*
character: *kharaktêr*
to charge into: *emballein*
choice: *hairesis*
choice-worthy: *hairetos*
to choose: *hairein*
clear: *tranês*
to come: *erkhesthai*
to compose: *graphein*
composite: *sunthetos*
concatenation: *heirmos*
concealment: *agnôstos*
conceptual content: *noêma*
configuration: *skhêmatismos*
configuration, entire: *suskhêmatismos*
to confirm: *kuroun*
conjecture: *huponoia*
to be consequent: *parepesthai*
constitution: *kataskeuê*
to contend: *katateinein*
cosmic: *kosmikos*

to count: *arithmein*
to create: *entekein*
to credit: *aitiasthai*
cycle: *periodos*

daimon: *daimon*
death: *thanatos*
declaration: *diagoreusis*
defence: *apologia*
degree: *moira*
degree of the zodiac: *hôra*
to deliver: *pherein*
demiurge: *dêmiourgos*
demiurgic: *dêmiourgikos*
to demonstrate: *apodeiknumnai*
to depend: *artasthai, exartan*
to be dependant: *exartasthai*
to deprive: *aphairein*
to determine: *aphorizesthai, horizein*
different: *diaphoros*
difficult: *duskolos, khalepos*
difficult to accomplish: *dusepiteuktos*
to direct: *agein*
to direct one's attention: *apoblepein*
to discover: *heuriskein*
discursive passage: *diexodos*
to discuss: *dialegesthai*
discussion: *dialexis*
disorderly: *ataktôs*
to display: *apophainein, endeiknunai*
divine: *theios*
to divine: *katamanteuein*
to do: *dran*
to do away with: *aphistanai*
dog: *kuôn*
downward journey: *kathodos*
dream: *onar*

earth: *gê*
effort: *spoudê*
Egyptian: *Aiguptios*
empty: *diakenos*
to encounter: *entunkhanein*
end: *telos*
to be endowed with form: *morphesthai*
to be enslaved: *douleuein*
to ensure: *paraphulattein*
entering: *eiskrisis*
entrance: *eisodos*
to entrust: *epitrepein*
to evade: *pheugein*
to be established: *keisthai*
eternal: *aidios*
evil: *kakos*
to be excellent: *aristeuein*
excess: *huperbolê*
executioner: *dêmios*
exotic: *barbarikos*
to expound: *apodidonai*

fair share: *moros*
to fall: *empiptein, katerkhesthai*
farmer: *geôrgikos*
to fasten: *exaptein*
fate: *heimarmenê, moira*
Fates: *Moirai*
fear: *phobos*
fearful: *periphobos*
feminine: *gunaikeios*
fiction: *plasmatôdês, pseudos, plasma*
fiery: *diapuros*
figure: *skhêma*
to be filled: *plêrousthai*
to fly up: *anaptênai*
to follow: *hepesthai*
following: *akolouthos*
folly: *atasthalia*
to force: *anangkazein, katanangkazein*
to forecast: *apotelein*
to forget utterly: *eklanthanesthai*
forgetfulness: *lêthê*
to form: *plattein*
to form a character: *êthopoiein*
formation of character: *êthopoiia*
fortune: *lêxis*
to frighten: *dedissesthai*
frightful: *phriktos*
fulfiller: *apoplêrôtês*

to generate: *gennân*
gift: *dosis*
to give in: *empiptein*
to go: *erkhesthai, ienai, pherein*
god: *theos*
good: *agathos, esthlos*
good manner: *kalôs*
grace: *kharis*
to grant: *didonai*
to grow lax: *luesthai*
guard: *prostatês*

habit: *epitêdeuma*

Hades: *Hadês*
hand: *kheir*
to hand down as tradition: *paradidonai*
haste: *epidromê*
to hate: *misein*
to have arrived: *hêkein*
to have in mind: *enthumein*
to have scientific knowledge: *epistenai*
having no form: *amorphôtos*
to hear: *akouein*
heaven: *ouranos*
hesitant to commit: *eulabês*
to hide: *kruptein, kruptesthai*
to hinder: *empodizein*
to honour: *atimazein*
horrible: *phrikôdês*
human: *anthrôpeios*
human being:*anthrôpos*
human reason: *logismos*
hypothesis: *hupothesis*

ignorance: *amathia*
to illustrate bad taste: *psukhreuesthai*
immense: *diôlugios*
immortal: *athanatos*
immortality: *athanasia*
impulse: *hormê*
in our power: *eph' hêmin*
incorporeal: *asômatos*
incredible: *paradoxos*
to indicate: *diasêmainein, endeiknunai*
individual: *idiotês*
inescapable: *anapodrastos*
inevitable: *aparaitêtos*
to be initiated: *teleisthai*
injustice: *adikia*
to inscribe: *epigraphein*
intellect: *nous*
intellection: *noêsis*
intention: *boulêma*
interpreter: *exêgêtês*
to intersect: *enallattesthai*
to introduce: *epeisagagein*
irreversible: *ametatropos*
issue: *logos*

Justice: *Dikê*

kind: *êpios*
kingly: *basilikos*
kingship: *basileia*
kinship: *sungeneia*
knowledge: *epistêmê*
last: *eskhatos*
law: *nomos*
to lead: *agein, apagein*
leading: *koruphaios*
to learn: *manthanein*
to leave: *apoleipein*
to be led around: *periagesthai*
licentious behaviour: *aselgeia*
life: *bios, zôê*
life-form: *diazêsis*
light: *phôs*
to be like: *eoikein*
likeness: *aphomoiôma*
to limit: *stenokhôrein*
lion: *leôn*
to listen: *akouein*
to live: *bioun*
to live out: *apopimplanai, diexagein, ektelein*
loss, to be at a: *aporein*
lot: *klêros*
to love: *agapan, philein*
lover: *erastês*
lucky: *epitukhôs*
luminous: *augoeidês*
lunar: *selênaios, selêniakos*

to make: *apergazesthai, poiein*
to make clear: *dêloun, emphanein*
to make mention: *mimnêskein*
to make subservient: *hupotattein*
male: *arrên*
maleficent: *kakopoios*
man: *anêr*
to manage: *dioikein*
to manipulate like a puppet: *neurospastein*
manner: *tropos*
masculine: *andreios*
mean: *meson*
meaning: *gnomê*
to meet: *apantan*
mercantile: *emporikos*
messenger (Er): *angelos*
middle: *mesos*
might: *sthenos*
to be mistaken: *aposphallesthai*
moderation: *mesotês*

month: *mên*
moon: *selênê*
mortal: *brotos*
mouth: *stomion*
to move: *kineisthai*
myth: *muthos*

nativity: *genesis*
nature: *phusis*
necessary: *anankaios*
to necessitate: *anangkazein, katanangkazein*
necessity: *anangkê*
nightingale: *aêdôn*
non-rational: *alogos*
nonsense: *phlauria*
not suited: *asummetros*
number: *arithmos*

oath: *horkos*
to object: *epikalein*
object of fear: *deima*
to be oblivious: *epilanthanesthai*
observation of the ascendant: *hôroskopia*
to obtain by lot: *klêrousthai*
one must flee: *pheukteon*
one-dimensional: *monotropos*
opposing: *ekpodôn*
order: *diakosmêsis, taxis*
outside: *exô, exôthen*
to own: *ktasthai*

pain: *algos*
paradigm: *paradeigma*
part: *meros*
to partake: *meteinai*
to participate: *metekhein*
particular: *kata meros*
to pass through: *diexerkhesthai*
passion: *pathos*
path: *hodos*
pathway: *diexodos*
peculiarity: *idiotês*
perfect: *teleos*
to perform: *dran*
philosopher: *philosophos*
philosophical: *philosophos*
place: *topos*
place of punishment: *kolastêrion*
plain: *pedion*
poet: *poiêtês*
pointless: *mataios*
to posit: *tithenai*
to position: *tattein*
position of leadership: *dêmagôgia*
position of power: *arkhê*
positive endowment: *eumoria*
to be possible: *endekhesthai*
power: *dunamis*
power of representation: *phantasia*
power to choose: *ethelousios*
preceding: *prooimios*
to prescribe: *diagoreuein*
to present: *proteinein, paradidonai*
priest: *hiereus*
to proceed: *khôrein*
proof: *apodeixis, pistis*
properly: *orthôs*
prophecy: *mantikê*
prophet: *prophêtês*
to provide: *porizein, parekhein*
providence: *pronoia*
to pursue: *metienai*

rainbow: *iris*
rank: *taxis*
to ratify: *epikuroun*
rational: *logikos*
reasonably: *eikotôs*
to receive: *paralambanein*
to receive by lot: *lankhanein*
receiving lots at same time: *sunklêros*
to reduce: *anagein*
to refer: *anapempein, anapherein*
to refer enigmatically: *ainissesthai*
to regulate: *dioikein*
to reject: *aposkeuazein*
relevant: *oikeios*
to remain: *menein*
to remind: *mimnêskein*
to render: *apotelein*
representative: *ekphantikos*
to reproach: *loidorein, oneidizein*
resembling: *prosemphorês*
to reveal: *deiknunai, ekphainein*
to revise: *diaskeuazein*
revolution: *periphora, peristrophê*
rhetorical: *logographikos*
as a kind of riddle: *ainigmatôdôs*
rightly: *kalôs*
rising: *anaphora, anatolê*

to be at risk: *kinduneuein*
rotation: *epistrophê*

sacred: *hieros*
scientific: *epistêmonikos*
second: *deuteros*
secret: *kruphios*
to see: *horan, idein, kathoran*
to seep into: *engignesthai*
self-determination: *autexousios*
sense of style: *mousa*
servant: *hupêretês*
to set up: *tithenai*
shame, to be without: *aperuthrian*
shameful: *aiskhros*
to show: *apophainein, deiknunai*
side-by-side: *parallêlos*
to signify: *sêmainein*
slave: *doulos*
sophistical: *sophistikos*
soul: *psukhê*
species: *eidos*
sphere: *sphaira*
to spin: *epinein*
to spin a destiny: *epiklôthein*
spindle: *atraktos*
to be stained: *khrainesthai*
star: *astêr*
starting point: *arkhê*
to steer: *kubernan*
to stop: *katalêgein, pauein*
story-line: *plasis tou muthou*
story-telling: *muthologia*
stylistically immature: *meirakiôdês*
to suffer: *pathein*
suffering: *pathêma*
to suppose: *hêgeisthai*
surprising: *thaumastos*
to suspect: *hupopteuein*
swan: *kuknos*

to take a stand: *istenai*
to take, receive: *lambanein*
tale: *diêgêma*
teacher: *didaskalos*
to tell stories: *muthologein*
text: *lexis*
theory: *theôria*
thief, to be a: *lêsteuein*
thing: *pragma*
thinking: *katanoêsis*
third: *triton*
three: *tria*
time: *diatribe, khronos*
tracking: *ikhneutikos*
tragic: *tragikos*
tragic poet: *tragôdiopoios*
training: *katataxis*
treatise: *logos, pragmateia*
true: *alêthinos*
truth: *alêtheia*
to try: *enkheirein*
tune: *mousa*
turning away: *paraklisis*
twelve: *dôdeka*
twentieth: *eikostos*
tyrant: *turannos*

unalterable: *aparabatos*
unchangeable: *akinêtos*
unconfined: *apolutos*
universe: *to pan*
unjust: *adikos*
to be unjust: *adikein*
unlawful: *athesmos*
unnecessary: *perittos*
upward journey: *anodos*
to use: *khrasthai*
to use an affectionate diminutive: *hupokorizesthai*
to use language: *phthengesthai*

vehicle: *okhêma*
vice: *kakia*
view: *ennoia*
to be one's view: *areskein*
virtue: *arête*

waking vision: *hupar*
to waste time: *diatribein*
watchdog: *oikophulax*
watcher: *epoptês*
wild: *agrios*
wise: *sophos*
without provision: *gumnos*
woman: *gunê*
word: *onoma*
world: *kosmos*
to write: *graphein*

zodiac: *zôidiakos*

Greek-English Index

Subject Index

Bibliography

Adamson, P. (2009). 'Plotinus on Astrology', *Oxford Studies in Ancient Philosophy* 35: 265-91.

Archer-Hind, R.D. (1888). *The Timaeus of Plato*. London, Macmillan.

Armstrong, A.H. (1966-88). *Plotinus Enneads*. Cambridge MA, Harvard University Press.

Atkinson, M. (1983). *Plotinus: Ennead V. 1. On the Three Principal Hypostases. A Commentary with Translation*. Oxford, Oxford University Press.

Aubry, G. (2008). 'Capacité et convenance: la notion d'*epitêdeiotês* dans la theorie porphyrienne de l'embryon', in L. Brisson, M.-H. Congourdeau and J.-L. Solère (2008): 139-56.

Babut, D. and Casevitz, M. (2004). *Plutarch Oeuvres Morales: Traités 70 et 71, Text, Translation and Commentary*. Paris, Les Belles Lettres.

Balme, D.M. (1992). *Aristotle. De Partibus Animalium I and De Generatione Animalium I*. Oxford, Oxford University Press.

Baltes, M. (1985/1999). Review of Deuse (1983), in *Göttingische Gelehrte Anzeigen* 237 (1985) 197-213. Reprinted in M. Baltes (1999).

Baltes, M. (1999). *Dianoêmata. Kleine Schriften zu Platon und zum Platonismus*. Stuttgart, Teubner.

Baltes, M. (2002). *Der Platonismus in der Antike*, vol. 6.2. Stuttgart, Bad Cannstatt.

Bertier, J. (1992). 'Médecine et philosophie dans l'*Ad Gaurum, sur la manière dont l'embryon s'anime*' in J.A. Lopez-Ferez (ed.), *Tratandos hipocráticos: actas del VIIe colloque international hippocratique*. Madrid: 635-45.

Beutler, R. (1953). 'Porphyrios (21)', *RE* XXII: 275-313.

Beutler, R. and Theiler, W. (1956-71). *Plotins Schriften*. Leipzig, Meiner.

Bien, C.G. (1997). *Erklärungen zur Entstehung von Mißbildungen im physiologischen und medizinischen Schriftum der Antike*. Stuttgart, Franz Steiner.

Blumenthal, H.J. (1971). *Plotinus' Psychology: His Doctrines of the Embodied Soul*. The Hague, Nijhoff.

Blumenthal, H.J. (1996). *Aristotle and Neoplatonism in Late Antiquity. Interpretations of De anima*. London, Duckworth.

Boll, F. (1894). *Studien über Claudius Ptolemäus*. Jahrbücher für klassische Philologie, Supplementband 21: 51-244.

Boll, F. (1903). *Sphaera*. Leipzig, Teubner.

Boll, F., Bezold, C. *et al.* (1966). *Sternglaube und Sterndeutung. Die Geschichte und das Wesen der Astrologie*. Darmstadt, Wissenschaftliche Buchgesellschaft.

Bouché-Leclercq, A. (1899). *L'Astrologie grecque*. Paris, Ernest Leroux.

Boudon-Millot, V. (2007), *Galien: Introduction générale; Sur l'ordre de ses propres livres; Sur ses propres livres; Que l'excellent médecin est aussi philosophe*. Paris, Les Belles Lettres.

Boudon-Millot, V. (2008). 'La naissance de la vie dans la théorie médicale et philosophique de Galien', in L. Brisson, M.-H. Congourdeau and J.-L. Solère (2008): 79-94.

Boudon-Millot, V. and Pietrobelli, A. (2005). 'Galien ressuscité: edition "princips" du text grec du *de Propiis Placitis*', *Revue des Études Grecques* 118: 168-213.
Boys-Stones, G. (2007). 'Physiognomy and Ancient Psychological Theory', in S. Swain (ed.), *Seing the Face, Seeing the Soul*. Oxford, Oxford University Press.
Brisson, L. (ed.) (2005). *Porphyre. Sentences*. Paris, Vrin.
Brisson, L., Congourdeau, M.-H. and Solère, J.-L. (eds) (2008). *L'Embryon. Formation et animation*. Paris, Vrin.
Brisson, L. and M. Patillon (1994). 'Longinus Platonicus Philosophus et Philologus, I. Longinus Philosophus', *ANRW II* 36(7): 5214-99.
Brisson, L. and Pradeau, J.-L. (2003). *Plotin. Traités 7-21*. Paris, Flammarion.
Carlier, J. (1998). 'L'après-mort selon Porphyre', in A. Charles-Saget (1998): 133-60.
Charles-Saget, A. (ed.) (1998). *Retour, repentir et constitution de soi*. Paris, Vrin.
Clark, G. (2000). *Porphyry: On Abstinence from Killing Animals*. London/Ithaca, Duckworth/Cornell University Press.
Congourdeau, M.-H. (2007). *L'embryon et son âme dans les sources grecques (VIe siècle av. J.-C. – Ve siècle apr. J.-C.)*. Paris, Association des amis du centre d'histoire et civilisation de Byzance.
Congourdeau, M.-H. (2008). 'La posterité byzantine de l'*Ad Gaurum*' in L. Brisson, M.-H. Congourdeau and J.-L. Solère (2008): 185-98.
De Lacy, P. (1984). *Galen. On the Doctrines of Hippocrates and Plato: Edition, translation and commentary*. Berlin, Akademie.
De Lacy, P. (1992). *Galen. On Semen*. Berlin, Akademie.
Dean-Jones, L. (1994). *Women's Bodies in Classical Greek Science*. Oxford, Oxford University Press.
Debru, A. and Garofalo, I. (2005). *Galien: Les os pour les débutants; L'anatomie des muscles*. Paris, Les Belles Lettres.
Deuse, W. (1983). *Untersuchungen zur mittelplatonischen und neuplatonischen Seelenlehre*. Mainz/Wiesbaden, Akademie der Wissenschaften und Literatur/Steiner.
Dillon, J.M. (1973). *Iamblichi Chalcidensis in Platonis dialogos commentariorum fragmenta*. Leiden, Brill.
Dillon, J.M. (1993). *Alcinous. The Handbook of Platonism*. Oxford, Clarendon.
Dodds, E.R. (1963). *Proclus. The Elements of Theology*. Oxford, Oxford University Press.
Dorandi, T. (2008). 'Pour une histoire du texte du traité *Ad Gaurum* attributé à Galien', in L. Brisson, M.-H. Congourdeau and J.-L. Solère (2008): 123-38.
Dörrie, H. (1959). *Porphyrios' 'Symmikta Zetemata'. Ihre Stellung im System und Geschichte des Neuplatonismus nebst einem Kommentar zu den Fragmenten*. München, Beck'sche Verlagsbuchhandlung.
Dörrie, H. (1976). *Platonica minora*. München, Fink.
Dörrie, H. and Baltes, M. (2002). *Der Platonismus in der Antike. Grundlagen – System – Entwicklung*, vol. 6. Stuttgart/Bad Cannstatt, Frommann/Holzboog.
Drews, U. (1993). *Taschenatlas Embryologie*. Stuttgart, Thieme.
Edelstein, L. (1967). *Ancient Medicine*. Baltimore & London, Johns Hopkins University Press.
Einarson, B. and Link, G.K.K. (1976-90). *Theophrastus. De Causis Plantarum*, 3 vols. Cambridge MA, Harvard University Press
Emilsson, E.K. (1988). *Plotinus on Sense-Perception: A Philosophical Study*. Cambridge, Cambridge University Press.
Erler, M. (1994). 'Epikur – Die Schule Epikurs – Lukrez', in H. Flashar (ed.),

Grundriß der Geschichte der Philosophie. Die Philosophie der Antike 4. Die hellenistische Philosophie, Part 1. Basel, Schwabe: 29-490.

Evans, J. (2004). 'The Astrologer's Apparatus: A Picture of Professional Practice in Greco-Roman Egypt', *Journal for the History of Astronomy* 35:1-44.

Farquharson, A.S.L. (1968). *The Meditations of the Emperor Marcus Antoninus*. Oxford, Oxford University Press.

Federspiel, M. (1998). 'Sur un emploi de sèmeion dans les mathématiques grecques', in G. Argoud et J.-Y. Guillaumin (eds), *Sciences exactes et sciences appliquées à Alexandrie (IIIe siècle av. J.-C.–Ier siècle ap. J.-C.), Actes du Colloque international de Saint-Étienne (6-8 juin 1996)*. Saint-Étienne, Université de Saint-Étienne: 55-72.

Festugière, A.J. (1944-54). *La révélation d'Hermes Trismégiste*. Paris, Lecoffre.

Festugière, A.J. (1970). *Proclus. Commentaire sur la République*. Paris, J. Vrin.

Fichtner, G. (1997). *Corpus Galenicum. Verzeichnis der galenischen und pseudogalenischen Schriften*. Tübingen, Institut für Geschichte der Medizin.

Finamore, J. and Dillon, J.M. (2002). *Iamblichus. De anima*. Leiden, Boston & Köln, Brill.

Frede, M. (1999). 'Stoic Epistemology', in K. Algra, J. Barnes, J. Mansfeld and M. Schofield (eds), *Cambridge History of Hellenistic* Philosophy. Cambridge, Cambridge University Press: 295-322.

Frommhold, K. (2004). *Bedeutung und Berechnung der Empfängnis in der Astrologie der Antike*. Münster, Aschendorff Verlag.

Gombosi, O.J. (1939). *Tonarten und Stimmungen der antiken Musik*. Copenhagen, E. Munksgaard.

Goulet-Cazé, M.-O. (1994). 'Chrysaorios', in R. Goulet (ed.), *Dictionnaire des philosophes antiques*, vol. 1. Paris, CNRS.

Gourevitch, D. (2006). 'Mynoide Mynas, un drôle de pistolet', in V. Boudon-Millot *et al.* (eds), *Ecdotica e Ricezione dei Testi medici greci*. Naples, M. d'Auria: 481-503.

Greenbaum, D.G. (2009). 'The Daimon in Hellenistic Astrology. Origins and Influence'. PhD thesis, University of London.

Gundel, H. (1972). 'Zodiakos', *RE* X.A. Berlin, Druckenmüller: 462-709.

Gundel, W. (1950). 'Dekane', *RE* S.VII. Stuttgart, Druckenmüller: 116-24.

Gundel, W. and Gundel, H.G. (1966). *Astrologoumena*. Wiesbaden, Steiner.

Hankinson, R.J. (ed.) (2008). *The Cambridge Companion to Galen*. Cambridge, Cambridge University Press.

Hanson, A.E. (2008). 'The Gradualist View of Fetal Development', in L. Brisson, M.-H. Congourdeau and J.-L. Solère (2008): 95-108.

Harder, R., Theiler, W. and Beutler, R. (1956-71) *Plotins Schriften*, 6 vols. Hamburg: Meiner.

Heilen, S. (2010). 'Ptolemy's Doctrine of the "Terms" and its Reception', in A. Jones (ed.), *Ptolemy in Perspective: Use and Criticism of his Work from Antiquity to the Nineteenth Century*. Archimedes 23. New York, Springer.

Helmreich, G. (ed.) (1907-9). *Galen. de Usu Partium*, 2 vols. Leipzig, Teubner.

Henry, P. and Schwyzer, H.-R. (1964-82). *Plotini Opera*. Oxford: Oxford University Press.

Housman, A.E. (1930). *M. Manilii Astronomicon liber quintus*. London, Richards Press.

Hübner, W. (1975). 'Das Horoskop der Christen (Zeno 1,38 L.)', *Vigiliae Christianae* 29: 120-37.

Hübner, W. (1990). '*Uranoscopus*. Der verstirnte Sterngucker', *Rheinisches Museum* NF 133: 264-74.
Hübner, W. (2001). 'Zur Verwendung und Umschreibung des Terminus *hôroskopos* in der astrologischen Lehrdichtung der Antike', *Mene* 1: 219-38.
Hübner, W. (2006). *Crater Liberi. Himmelspforten und Tierkreis*. Munich, Bayerische Akademie der Wissenschaften – philosophisch-historische Klasse (Heft 3).
Jouanna, J. (2008). 'La postérité de l'embryologie d'Hippocrate dans deux traités pseudo-hippocratiques de la médicine tardive', in L. Brisson, M.-H. Congourdeau and J.-L. Solère (2008): 15-42.
Jurisch, U. (1991). 'Grundfragen der Embryonalentwicklung aus der Sicht eines Neuplatonikers: Übersetzung und Bearbeitung der Schrift des Porphyrios über die Beseelung der Embryonen Medicine', PhD thesis, Friedrich Alexander Universität Erlangen-Nürnberg.
Kalbfleisch, K. (1895). Die neuplatonische, fälschlich dem Galen zugeschriebene Schrift *Pros Gauron peri tou pôs empsukhoutai ta embrua. Aus dem Anhang zu den Abhandlungen der Königl. Preuss. Akademie der Wissenschaften zu Berlin vom Jahre 1895*. Berlin, Verlag der königl. Akademie der Wissenschaften.
Kirk, G.S., Raven, J.E. and Schofield, M. (1983). *The Presocratic Philosophers*. Cambridge, Cambridge University Press.
Kobusch, T. (1976). *Studien zur Philosophie des Hierokles von Alexandrien*. Munich, Berchman.
Kroll, W. (1899-1901). *In Platonis Rempublicam commentarii*. Leipzig, Teubner.
Kroll, W. (1924). 'Kyraniden', *RE* XII.1 (Stuttgart), 127,19-134,31.
Lenz, H.O. (1856). *Botanik der alten Griechen und Römer*. Wiesbaden: Sändig.
Lesky, E. (1951). *Die Zeugungs- und Vererbungslehren der Antike und ihr Nachwirken*. Wiesbaden, Verlag der Wissenschaften und der Literatur.
Lesky, E. and Waszink, J.H. (1959). 'Embryologie', *Reallexicon für Antike und Christentum* 4: 1233-8.
Lienau, C. (1973). *Hippocrates über Nachempfängnis, Geburtshilfe und Schwangerschaftsleiden*. Berlin, Akademie.
Limburg, K. (1975). 'Porphyrios, Die Beseelung der Embryonen', PhD thesis, University of Cologne.
Lloyd, A.C. (1976). 'The Principle that the Cause is Greater than its Effect', *Phronesis* 21: 146-56.
Lloyd, G.E.R. (1996). *Aristotelian Explorations*. Cambridge, Cambridge University Press.
Long, A.A. (1982). 'Astrology: Arguments Pro and Contra', in J. Barnes, J. Brunschwig, M. Burnyeat and M. Schofield (eds), *Science and Speculation: Studies in Hellenistic Theory and Practice*. Cambridge, Cambridge University Press: 165-92.
Long, A.A. and Sedley, D.N. (1987). *The Hellenistic Philosophers*. Cambridge, Cambridge University Press.
Longrigg, J. (1998). *Greek Medicine from the Heroic to the Hellenistic Age: A Sourcebook*. London, Routledge.
Lonie, I.M. (1981). *The Hippocratic Treatises 'On Generation', 'On the Nature of the Child' and 'Diseases IV'*. Berlin & New York, Walter de Gruyter.
Majercik, R. (1989). *The Chaldean Oracles*. Leiden, Brill.
Mau, J. (1960). *Galen. Einführung in die Logik*. Berlin, Akademie.
May, M.T. (1968). *Galen. On the Usefulness of the Parts of the Body. (Translation with notes)*. Ithaca, New York, Cornell University Press.

Mayhew, R. (2004). *The Female in Aristotle's Biology*. Chicago, Chicago University Press.
Mohr, H. and Schopfer, P. (1995) *Plant Physiology*, tr. Gudrun and David W. Lawlor, 4th edn. Berlin, New York & Heidelberg, Springer.
Nickel, D. (1989). *Untersuchungen zur Embryologie Galens*. Berlin, Akademie.
Nickel, D. (2001). *Galen. De foetuum formatione (Über die Ausformung der Keimlinge)*. Corpus Medicorum Graecorum 5 3,3. Berlin, Akademie.
Nutton, V. (1999). *Galen. On My Own Opinions (Text, Translation and Commentary)*. Corpus Medicorum Graecorum 5 3,2. Berlin, Akademie.
O'Brien, D. (1977). 'Le volontaire et la nécessité: reflexions sur la descente de l'âme dans la philosophie de Plotin', *Revue philosophique de la France et de l'étranger* 167: 401-22.
O'Daly, G.J.P. (1973). *Plotinus' Philosophy of the Self*. Shannon, Irish University Press.
Omont, H. (1916). 'Minoide Mynas et ses missions en Orient (1840-1855)', *Mémoires de l'Académie des Inscriptions et Belles-Lettres* 40: 337-421.
Pépin, J. (1974). 'La fortune du *De antro Nympharum* de Porphyre en Occident', in *Plotino e il Neoplatonismo in Oriente e Occidente*. Rome, Accademia Nazionale dei Lincei: 527-36.
Pfister, F. (1938). 'Pflanzenaberglaube', *RE* 19(2): 1446-56.
Pohlenz, M. (1948). *Die Stoa. Geschichte einer geistigen Bewegung*, 2 vols. Göttingen, Vandenhoeck & Ruprecht.
Remes, P. (2007). *Plotinus on Self: The Philosophy of the 'We'*. Oxford, Oxford University Press.
Risch, F.X. (2007). 'Die porphyrische Schrift *ad Gaurum* als *hieros logos*', *Zeitschrift für antikes Christentum* 10: 260-75.
Rocca, J. (2003). *Galen on the Brain: Anatomical Knowledge and Physiological Speculation in the Second Century AD*. Leiden, Brill.
Sadler, T.W. (2004). *Langman's Medical Embryology*. Philadelphia PA, Lippincott, Williams and Wilkins.
Saffrey, H.D. and Westerink, L.G. (1968-1997). *Proclus. Théologie Platonicienne*, 6 vols. Paris, Les Belles Lettres.
Schibli, H.S. (1990). *Pherekydes of Syros*. Oxford, Oxford University Press.
Schulze, S. (2006). *Kurzlehrbuch Embryologie*. Munich & Jena, Urban & Fischer.
Siegel, R.E. (1970). *Galen on Sense Perception*. Basel, S. Karger.
Singer, P.N. (1997). *Galen. Selected Works*. Oxford, Oxford University Press.
Sleeman, J.H. and Pollet, G. (1980). *Lexicon Plotinianum*. Leiden, Leuven University Press.
Smith, A. (1974). *Porphyry's Place in the Neoplatonic Tradition: A Study in Post-Plotinian Neoplatonism*. The Hague, Nijhoff.
Smith, A. (1993). *Porphyrii Philosophii Fragmenta*. Stuttgart & Leipzig, Teubner.
Sorabji, R. (2000). *Emotions and Peace of Mind: From Stoic Agitation to Christian Temptation*. Oxford, Oxford University Press.
Sorabji, R.S. (2004). *The Philosophy of the Commentators, 200-600 AD. A Sourcebook*. London/Ithaca, Duckworth/Cornell University Press.
Stahl, W. H. (1952). *Commentary on the Dream of Scipio by Macrobius*. New York, Columbia University Press.
Strange, S. (1992). *Porphyry. On Aristotle Categories*. London/Ithaca, Duckworth/Cornell University Press.
Theiler, W. (1930). *Die Vorbereitung des Neuplatonismus*. Berlin, Weidmann.

Theiler, W. (1966). 'Ammonios und Porphyrios', in *Porphyre (Entretiens XII)*. Vandoeuvres & Genève, Foundation Hardt: 85-119.
Theiler, W. (1966). *Forschungen zum Neuplatonismus*. Berlin, De Gruyter.
Tieleman, T. (2003). 'Galen's Psychology', in *Galen et la philosophie*. Vandoeuvres & Genève, Fondation Hardt: 131-69.
Tornau, C. (1998). *Plotin. Enneaden VI 4-5 [22-23]. Ein Kommentar*. Stuttgart & Leipzig, Teubner.
van den Berg, R.M. (2001). *Proclus' Hymns: Essays, Translations, Commentary*. Leiden, Boston & Köln, Brill.
van der Eijk, P. (1994). *Aristoteles. De insomniis. De divinatione per somnum*. Berlin, Akademie.
van der Eijk, P. (2005). *Medicine and Philosophy in Classical Antiquity*. Cambridge, Cambridge University Press.
Vetter, W. (1933). 'Musik', *RE* XVI(1): 843.12ff.
von Straaten, M. (1975). 'On Plotinus IV.7 [2].8[3]', in J. Mansfeld and L.M. de Rijk (eds), *Kephalaion*. Assen, The Netherlands: 164-70.
Waszink, J.H. (1947). *Tertullian. De Anima*. Amsterdam, Meulenhoff.
Waszink, J.H. (1954). 'Beseelung', *Reallexicon für Antike und Christentum* 2: 176-83.
West, M.L. (1971). *Early Greek Philosophy and the Orient*. Oxford, Clarendon.
Westerink, L.G. (1971). 'Ein astrologisches Kolleg aus dem Jahr 564', *Byzantische Zeitschrift* 64: 6-21.
Whittaker, J. and Louis, P. (1990). *Alcinoos, Enseignement des doctrine de Platon*. Paris, Les Belles Lettres.
Wilberding, J. (2005). '"Creeping Spatiality": The Location of *Nous* in Plotinus' Universe', *Phronesis* 50.4: 315-34.
Wilberding, J. (2006). *Plotinus' Cosmology. A Study of Ennead II.1 (40). Text, Translation, and Commentary*. Oxford, Oxford University Press.
Wilberding, J. (2008). 'Porphyry and Plotinus on the Seed', *Phronesis* 53.4-5: 406-32.
Wilberding, J. (forthcoming). 'The Myth of Er and the Problem of Constitutive Luck', in A. Sheppard (ed.) *The Reception of Plato's Republic*. London, BICS.
Wilberding, J. (forthcoming). 'The Secret of Sentient Vegetative Life in Galen', in P. Adamson and J. Wilberding (eds), *Philosophical Themes in Galen*. London, BICS.
Zambon, M. (2002). *Porphyre et le Moyen-Platonisme*. Paris, Vrin.
Zeyl, D.J. (2000). *Plato. Timaeus*. Indianapolis & Cambridge, Hackett.

Lightning Source UK Ltd.
Milton Keynes UK
UKOW04f0407030514

231051UK00001B/46/P

9 781472 557872